THE GREAT DISRUPTION

THE GREAT DISRUPTION

COMPETING AND SURVIVING IN THE SECOND WAVE OF THE INDUSTRIAL REVOLUTION

RICK SMITH with **MITCH FREE**

THOMAS DUNNE BOOKS ✷ ST. MARTIN'S PRESS   NEW YORK

To Lori Smith, mother of Arden, Adam, and

Laine, and Shirene Free, mother of Mia

THOMAS DUNNE BOOKS.
An imprint of St. Martin's Press.

www.thomasdunnebooks.com
www.stmartins.com

Designed by Richard Oriolo

Library of Congress Cataloging-in-Publication Data

Names: Smith, Rick, 1967– author. | Free, Mitch, author.
Title: The great disruption: competing and surviving in the second
    wave of the industrial revolution / Rick Smith with Mitch Free.
Description: New York: St. Martin's Press, [2016]
Identifiers: LCCN 2016017019 | ISBN 9781250091420 (hardcover) |
    ISBN 9781250091437 (e-book)
Subjects: LCSH: Technological innovations—Economic aspects—
    History. | Industries—Technological innovations. | Three-
    dimensional printing. | Strategic planning.
Classification: LCC HC79.T4 S65 2016 | DDC 338/.064—dc23
LC record available at https://lccn.loc.gov/2016017019

Our books may be purchased in bulk for promotional, educational,
or business use. Please contact your local bookseller or the
Macmillan Corporate and Premium Sales Department at 1-800-221-
7945, extension 5442, or by e-mail at MacmillanSpecialMarkets@
macmillan.com.

First Edition: October 2016

10 9 8 7 6 5 4 3 2 1

CONTENTS

DIGITIZATION AND THE INDUSTRIAL 3D PRINTING REVOLUTION: FROM "SHAPES" TO "INDUSTRIAL PARTS"

By Bernd Leukert,
Member of the Executive Board, SAP

IT MAY LOOK LIKE A relatively new invention, but the concept of 3D printing celebrated its thirty-fifth anniversary this year. In 1981, Hideo Kodama of Nagoya Municipal Industrial Research Institute published a paper describing a functional rapid proto-typing system, "printing" physical objects in layers. Seven years later, fused deposition modeling (FDM) was invented—a tech-nique that allows designers to create 3D models using digital data that can then be used to produce a tangible object.

While these technologies were by no means perfect, their potential was undeniable. But they also came with a hefty price tag. It wasn't until 2007 that the first 3D printer under $5,000 was introduced. As a consequence, the market went into consumer frenzy and a rapidly evolving industry fueled our dreams of having 3D printers in every household.

Nearly ten years later, this compelling consumer vision has still not been realized, remaining in the realm of the hobbyist. But, over the last five years, industrial 3D printing has significantly expanded and matured. The transition from prototypes and trinkets to fully on-demand manufacturing is here to stay, and is beginning to transform entire business models.

The reliability of industrial 3D printing today gives any manufacturing company, no matter the size or industry, the opportunity to consider 3D printing as an integral part of their production line capabilities. The technology is finally delivering on its original promise.

3D printing's leap forward from "printing shapes" to "manufacturing parts" clearly marks a transformative moment in the history of manufacturing. 3D printing can now tackle industrial scale problems for companies both large and small, including automation, efficient production and new ways to manage inventories and extended supply chains.

According to a recently published report by the U.S. Department of Commerce[1], the total manufacturing and trade inventories in the United States are estimated at $1,807.1 trillion.

The supply chain implications of reducing the U.S. inventories levels even by 1 percent alone are astonishing—$18 billion. The

1. See: www.census.gov/mtis/www/data/pdf/mtis_current.pdf

impact on the global economy, global supply chain, and even global climate are even greater.

But it does not stop there. 3D printing also allows companies to transform their business models and offer new digital services such as mass personalization, local manufacturing, same day delivery, to mention just a few examples.

That is why SAP is taking a leadership role, collaborating with prominent companies across manufacturing and extended supply chain to accelerate the transformation and digitization of manufacturing globally.

3D printing on a global scale has arrived. Brace yourself for the next Industrial Revolution—I wish you an inspiring read.

"This Changes Everything!"

THE TIME IS EARLY 2015. The place: a large boardroom in the Minneapolis headquarters of 3M, a global manufacturing company with annual sales of $30 billion.

A few months earlier, Mitch Free and I had decided to start a new venture in 3D printing. Mitch has a deep manufacturing background—he founded the global industrial marketplace MFG.com. I am what you could call a serial entrepreneur, having started businesses in several industries. We had both become fascinated with the emergence of industrial 3D printing. Our hunch was that the field already had some interest from Fortune 1000 companies. But neither of us knew for sure what we were venturing into until that day at 3M. To be honest, we were a bit anxious.

We introduced ourselves to Hector Dalton, 3M's executive vice president in charge of manufacturing and supply chain, and other company executives. The meeting paused in an initial strained silence, so I tried to get down to business. "Are there any specific areas where you believe that 3D printing will have a significant impact on 3M's business?"

Hector looked around the room, but no one said a word. Then Hector slowly leaned across the table toward us, arching his arms instinctively as if he were about to spring forward.

"There isn't *one* area. This changes everything! We are about to witness a disruption of historic proportions."

A historic disruption?

At that moment I was struck with a flash of insight. We commonly think about the Industrial Revolution as if it were a binary switch. There was a before and an after. All of us live *after* the historic change, in a world safely and firmly set in the modern era. *This IS the future!*

But what if that's not right? What if the wrenching transformation of industry is not finished? What if everything was about to be flipped upside down all over again?

My mind was spinning, but my hand wrote down three words in a notebook, as if I was whispering them to myself.

The Great Disruption!

Fast-forward six months past that meeting. Mitch and I had interacted with dozens of companies from many different industries.

They all had their own uses for 3D printing, but one thing had become clear. Our hunch about the size and scope of this historic disruption was now a firm belief. We, along with a growing number of leaders around the world, are convinced that 3D printing will change the way things are produced in this century more than the Industrial Revolution did over the past three hundred years. The Second Wave of the Industrial Revolution is coming, and the impacts are already beginning to wash up upon our feet. It may seem premature to make such claims. It is certainly bold. But it is not without cause.

Here's the quick explanation of how the technology works. Picture your current home or office document printer, but with two big differences. First, instead of printing with ink, a 3D printer uses plastics, ceramics, or even metal. Second, the printer doesn't do all its work in one pass and then move on to the next page. Instead a 3D printer continues going back and forth hundreds or even thousands of times. During each pass, it creates incredibly thin, precise layers on top of each other until you have a fully formed three-dimensional object.

Here's the amazing part: 3D printing, also known as additive manufacturing, didn't even exist until the late 1980s. Initially it was just making low-quality objects with plastic. Many people still associate the technology with novelty items. They are missing the disruption. Today, for example, General Electric is already 3D printing critical metal parts for jet engines. These components offer game-changing improvements in performance and weight, and cannot be made any other way. As the technology becomes cost competitive with traditional manufacturing, every other industry—from shoes to cars to foods and medicine, and even housing—has the potential to similarly benefit. Which raises a key question: What are the limits of this disruption?

The truth is that we don't know exactly. We have seen that 3D printing is massively disruptive by itself. But as we will explore in the pages ahead, its full impact will likely result from a synthesis with other emerging technologies. The potential for combining 3D printed manufacturing with artificial intelligence, hyper-customization, computer-generated design, and the Internet of Things is mind-boggling. Also, don't expect 3D printing's disruption to happen all at once. Many of the dramatic changes we discuss in this book will take years, in some cases even decades, to play out.

But let's be very clear: Behind the curtain, thousands of top-level executives are already aggressively preparing for this massive shift. Like these people, we are convinced that anyone looking to prosper or even survive in this soon-to-be reconfigured business world needs to begin understanding 3D printing and its potential right now.

The Great Disruption is real, and it is inevitable. This is not a matter of *if* but *how soon*. That's why we have written this book, not to dazzle you with gee-whiz technology that often seems ripped from the pages of science fiction (although that's part of the story) but to convince you that 3D printing will be enormously disruptive to *your* industry in *your* lifetime.

Throughout the book we will explain why the relevant question is no longer whether historic, sweeping change is coming. Instead, you should ask: "What can I and my organization be doing right now to prepare for this emerging future so that I will not only be able to survive, but thrive?"

1

PARADIGM PLUS

The future ain't what it used to be.

—YOGI BERRA

FOR GE, THE ICONIC GLOBAL industrial company founded by Thomas Edison, the Great Disruption began to emerge in late 2013. It came from the most unlikely of messengers: a young man working shoeless in front of twinned computer screens in a cramped room just outside of Jakarta, Indonesia. Arie Kurniawan had recently noticed on an engineering website that GE's aviation division was hosting an "Open Innovation Challenge" to redesign a critical component of an airplane. The competition sounded a bit wonky: The winner would create the best new design for a relatively simple-looking bracket. But this was no ordinary part. The bracket attached directly to an eight-thousand-pound jet engine and was critical in securing the engine to the wing. The part that GE was redesigning, one that had been in use for decades, was heavy and clunky but, as you would expect, incredibly strong and reliable.

It was late at night, and Arie had been working tirelessly on the project. His approach was not just to find improvements to this critical component, but to completely reimagine it. As dawn broke, Arie hit the send button, submitting his entry. He was satisfied with his efforts—as much as any engineer ever is—but he tried to keep his expectations realistic. In fact, he wondered if his submission would ever get more than a cursory glance from the veteran GE engineers on the selection committee.

A few months later, to Arie's complete surprise, General Electric chose his design over nearly a thousand other submissions. But the shock waves from his success spread far beyond his tiny room in Indonesia. First of all, it turned out that Arie had no experience whatsoever with industrial manufacturing—the design and production of large, highly durable machinery. In fact, his only formal design education had come a few years earlier in his provincial town, at a vocational high school with the slogan "Prepare faithful graduates who can compete in the global world." He had spent the last few years designing gloves and computer stands, not high-stress industrial equipment. But his lack of complex manufacturing experience didn't stop him from beating some very tough competition. One submission had come from a Swedish Ph.D. who worked for Saab and General Motors. Another was developed by a British stress engineer at Airbus, the world's largest producer of commercial aircraft.

At left, GE's old engine bracket. At right, Arie Kurniawan's replacement, one-sixth the weight of its predecessor and printed as a single unit using 3D technology. SOURCE: GE

But that's not all that was surprising about Arie's story. His novel design was enabled by new industrial 3D printing technology. Up until that point, every other engine bracket ever used— like many other of the hundreds of critical pieces that make up a jet—had been produced through conventional manufacturing. These techniques included pouring liquid metal into the hollow center of a mold, or bending and shaping molten ore. Yet the quirky-looking bracket Arie designed in his small office worked perfectly. Printed by General Electric with technology that is still in its early years, the bracket passed every one of the rigorous tests for durability, stress, and reliability.

And it weighed 83 percent less than the part it would replace.

Meanwhile, halfway around the world, a GE executive named David Joyce walked down a hallway toward a 3D printer. The machine was a refrigerator-sized box accessed by a door that swings open from the side. Inside, lasers were meticulously fusing together metal powder into an industrial-strength object, "printing" it layer by layer. David had spent almost his entire career, more than three decades, working for GE's aviation division. During that time he had tirelessly manipulated engineering designs, chasing incremental improvements in productivity and efficiency. This is tedious, relentless work. Even successes can be measured in the hundredths of a percent.

But that was before 3D printing handed GE a whole new range of tools. Now, his engineers could design components with complicated geometries that were previously impossible to make. These printers can essentially build a physical version of anything that can be imagined on a computer. This includes a huge number of objects that would be impossible to create using any other manufacturing method. 3D printers, for example, can drill the proverbial "curved hole" in a metal block—or print objects that exactly replicate that effect. They can create a seamless hollow steel ball, or one that contains another empty steel ball, which contains another steel ball, and on and on.

Curved holes and spheres within spheres are cool, but they are basically the manufacturing equivalent of party tricks. David's team was manufacturing a sophisticated industrial component with real-world use—and huge implications. The 3D printer stopped, and David peered inside at GE's brand-new fuel-injection nozzle for a jet engine.

The first amazing thing about this new part is simply that it was created as one piece. The previous nozzle had twenty-one separate parts, each of which needed to be produced by parts suppliers, shipped to a central location, and then assembled. Second, it was not created on a noisy factory floor, but in a laboratory that could just as easily have been in Nigeria or on the fifth floor of a Manhattan apartment building.

Its improvements in functionality were also massive. The new fuel-injection system was five times stronger, five times more durable, and 25 percent lighter than the old part. And it increased fuel efficiency by an astonishing 15 percent!

Taken together, these innovations will result in fuel cost savings of over $1 million annually. *For every single airplane that uses the new system.*

Reports of these two separate events quickly spread around General Electric, all the way to the office of the CEO. Certainly no one expected these parts to have an immediate impact on the company's overall financials, but the implications of these two events were disarmingly clear:

- **If a twenty-something with no training in industrial manufacturing could outdesign a leading multinational company stocked with top-flight engineers, what were the implications for the current global workforce?**

- **If these two critical parts could be redesigned with such massive improvements, what were the possibilities for the company's other tens of millions of parts?**

- **If twenty-one parts could be produced as one, what did this mean for the future of GE's long-standing parts producers?**

- **If this new technology allowed manufacturing to be cost effectively reshored to the United States, what could this mean for global supply chains?**

But perhaps most importantly, what if these new technologies could be used to redesign not only a few parts, but an entire airplane? Could we envision reducing the entire weight of a plane by 5 percent, 10 percent, *even 20 percent*? Outcomes like these would not simply result in a financial uplift for companies like GE. It would change the economics of an entire industry! *Indeed, it would change every industry.*

HIGH COMPLEXITY, LOW VOLUME, INFINITE CHANGE

GE executives aren't the only ones who are quickly beginning to understand what 3D printing will mean for the future. This technology is beginning to transform not just how companies make things, but how they design, ship, and warehouse them. Simultaneously, 3D printing is unleashing a flood of product innovation. The technology's impact even extends to company cultures, creating more collaborative work environments.

All this is possible because of 3D printing's two unique characteristics. First, the technology has an unparalleled capacity for high complexity manufacturing. If it can be created on a digital 3D modeling program, it can be made into a physical object. This is not the case with traditional production tools. Manufacturing GE's new fuel nozzle, for example, would be impossible without 3D printing.

Second, 3D printing enables low-volume, customized manufacturing, something that is also not economically feasible using today's industrial processes. Mass production dominates conventional manufacturing for one reason: It produces objects at very low unit cost. We all reap the benefits of this when we buy incredibly cheap socks or televisions or microwaves. But mass production also has several drawbacks. First, it is only profitable at scale. A company has to make and sell an exceedingly large number of the same product before it makes money. Mass production also requires a large initial investment. Companies will often invest tens of thousands of dollars, sometimes millions, before they can begin to produce the first of thousands of cheap, identical objects. Finally, mass production leaves no room for variation in the design. You do get cheap tablets. You don't get to customize them.

With 3D printing, all that changes. First of all, the costs of production are not front-weighted. This means that you don't have to make tens of thousands of the same thing before hitting profitability. In fact there are essentially no savings for making lots of identical products. Sure, 3D printed objects cost more per unit than mass-produced ones. But because unit costs are not tied to the number of pieces produced, you are also not penalized for producing in very small quantities. For example, if GE wanted to experiment with engine bracket design, they could print three different designs for roughly the same cost as printing three identical ones.

Bottom line: 3D printing technology is not just an interesting new way of making things. It's more than a paradigm shift in manufacturing. Ultimately, 3D printing technology is the Second Wave of the Industrial Revolution. It stands to sweep away a quarter millennium of manufacturing evolution. That's not a prediction, by the way; it's a pattern.

Think about it: All great technological transformations begin with mass production, whether it's the mass production of books (via Gutenberg's printing press), cars (Henry Ford's assembly line), durable appliances (Westinghouse dishwashers), advanced electronics (iPhones, iPads, etc.), or hundreds of other examples. Mass production creates a material bounty at prices low enough that most people can afford them.

But while cheap mass-produced items are what the masses get, customization is what we really want. There are very basic human reasons for this. First of all, mass-produced items target the average consumer. But none of us believes we are average. Given the chance at the same price, we will always choose something made uniquely for ourselves over something made generically for everyone. Second, customization also enables a full range of applications, which is in itself a powerful recognition of the full complexity of our individual human wants and needs.

BESPOKE BODIES

Medical researchers and suppliers were among the first to explore the technology's possibilities, and the early results of this innovation are astonishing. 3D technology allows surgeons and researchers to repair and replicate the human body in ways that were unimaginable even a decade ago. Again, the key is complexity and customization.

Take knee replacements, a procedure undergone by hundreds of thousands of people every year. Typically, the operation goes something like this. The surgeon slices open your knee and pins back the skin around it. An assistant has a few different sizes of replacements for the diseased or damaged portion of the knee. The surgeon holds the parts up to your knee and picks the best match. Then the surgeon puts the replacement in your knee and makes it fit as precisely as possible. You're sewn up and sent to physical therapy.

This is an approach riddled with problems. No off-the-shelf replacement can fit your body perfectly. The implant's lack of geometrical precision is compounded by the sophisticated role knees play in the body. Knees are your largest joint, are more complex than shoulders, and are weight-bearing. Many replacements are such poor fits that they don't support the body properly or they grind against other parts of the knee and leg. As a result, many patients experience postsurgical pain and more than a few have to go under the knife again, hoping for a better fit.

3D printing offers a new answer to these problems. In this scenario, the surgeon scans your knee and then prints out an exact replica—not the closest fit but a perfect copy of the part being replaced. These 3D printed replacements are inserted just like the pre-sized versions. But the postsurgical experience is significantly better. Patients have shorter hospital recovery periods, less pain, and far better initial movement than with the off-the-shelf replacements.

3D printed replacements for other body parts have had similar successes. Again, the technology creates exact geometrical replacements in sensitive areas cost-effectively. Among them are custom-fit parts of skulls to repair puncture wounds in the head. Additionally, the 3D part can be manufactured with intricate channels, encouraging integration with our bodies. The complex grooves encourage the bone to grow around the 3D printed part. In other cases, 3D printing's capacity for extreme design complexity is creating previously impossible medical supplies. For example, BASF is printing new skin for burn victims instead of relying on grafts from other parts of the body. Before the emergence of 3D printing, manufacturing these medical innovations was either completely impractical or flat out impossible.

In some cases, these made-to-order 3D printed body parts have a further advantage. They are not only cost-effective, but substantially cheaper than traditionally manufactured ones. In 2014, researchers at the University of Central Florida printed a prosthetic for a six-year-old boy born without a right arm below the elbow. His family's medical insurance would not pay for a traditionally manufactured one, which can cost up to $40,000. This is also a recurring expense since the arm will need to be replaced repeatedly over the next decade as the boy's body grows. The 3D printed prosthetic cost only $350.

Amazing, right? But then something even more surprising happened. The researchers put their models online in open-source files with the hope that others would improve on the design. This broke all the rules of traditional manufacturing. No medical supply company would ever make their designs free. A year later, a hobbyist heard about a seven-year-old boy who needed an arm. He wanted to make something the kid would be excited to wear, so he specially designed an arm that looked like the prosthesis used by *Star Wars*' Luke Skywalker. That designer model was even cheaper, costing just $300. The recipient was presented with his complimentary new arm by a parade of imperial storm troopers and other *Star Wars* characters.

In just a little over a decade, 3D printing has dramatically improved critical areas of medical care. Future applications are nearly limitless and will improve—or save—the lives of hundreds of millions of patients. True, medicine is in some ways tailor-made for early adoption of 3D printing. The industry has high product development budgets and overall margins, and there's nothing like a new *Star Wars*–themed prosthetic to bring out the news teams and generate publicity. But the two novel manufacturing characteristics behind these miracles—design freedom and low-to-no-cost customization—are much more broadly applicable. A technology that can print extremely complex, customized products on a massive scale will find a place in any industry.

ON-DEMAND UNIVERSE

In the late 1940s, television was beginning to supplant radio as the dominant form of mass entertainment. Hoping to cash in on the medium's popularity, an appliance dealer named John Walsonavich began loading his pickup truck with televisions and driving prospective customers up the steep mountains surrounding Mahanoy City, a town in the eastern coal region of Pennsylvania. At the top, Walsonavich hooked the televisions up to his antenna. This extra boost was the only way to get clear reception of the three Philadelphia network stations. Sometimes he made a sale, but his clientele was strictly limited to people who lived higher up the hillside. Nobody in lower-lying Mahanoy City got reception.

Then one day, frustrated with the constant sales trips, Walsonavich bought a long coil of wire from an army surplus store, connected it to his mountaintop antenna and ran the wire all the way down to his appliance shop. He hooked up the wire to three televisions, one for each network, and placed the sets in the front window of his shop. Television was still new enough that most people in Mahanoy City had never owned or even seen one. Crowds gathered on the street. Soon residents were asking Walsonavich to run the cable to their houses in the town. He charged a flat rate for connection but waived the monthly $2 fee for a year if the subscriber bought a TV from him. Walsonavich had created the world's first cable television network.

In the 1970s, Walsonavich expanded his company, now named Service Electric Cable TV, and broke new ground once again. Service Electric became the first affiliate for HBO, showing a New York Rangers hockey game. Over the next two decades, as stations like WTBS, CNN, and MTV appeared, the cable television that Walsonavich had pioneered began stealing viewers from traditional networks by offering a wider variety of options. By 2000, cable televisions had over 68 million paid subscribers—roughly half the population of the entire country in the 1940s, when Walsonavich was still driving customers to his mountaintop showroom. But 2000 also turned out to be the peak year for cable television. The service began a steady decline. Today there are about 54 million subscribers—or about as many as there were in 1991.

Why did this happen? The answer is rooted in the qualitatively distinct advantage of mass production versus radical customization.

Over the past twenty years, two major changes have radically altered televised entertainment. Initially, satellite-based networks like Dish and DirecTV stole some of cable's customers. This was just a bump in the road. In recent years, those services' viewership has also begun to shrink.

The more pernicious and existential threat to both cable and satellite television is the Internet. Over the past ten years, online services like Netflix and Hulu have allowed viewers to watch an enormous store of material, from *Scandal* to a cricket match to a Mexican telenovela to a classic movie like *Gentlemen Prefer Blondes*. Because of its massive variety of content it is tempting to think of the Internet as a kind of supersized version of cable. But Comcast and DirecTV aren't bleeding viewers every year just because the Internet offers a wider variety of content. Their audience is slipping away because the Internet allows viewers to customize their entertainment options. Citing Nielsen viewership trends, YouTube's CBO Robert Kyncl even predicts that digital video will overtake television by the end of the decade.

The fundamental difference between cable television's expansion in the 1970s and the growth of Internet viewership today is one of mass production versus customization. Cable became popular because it gave viewers more stuff. For a relatively low price, you got an enormous supply of additional options that weren't offered by the three national networks and a handful of local stations. Beyond doubt, this represents a vast quantitative difference *but* a minimal qualitative one. The additional content was an enormous mountain of standardized material produced for a mass audience. Cable networks multiplied the number of cop shows, dramas, situational comedies, cooking programs, and nature documentaries. But even with more stuff, there wasn't much more variability. Each program was still designed for the average viewer of that genre. There may be 150 channels today but—with cable—even niche networks still have to generate a mass-market viewership to survive. And mass-market still means (and always will mean) a huge clump of people watching the same show at the same time.

Enter the Internet. The Internet offers a qualitatively different experience from either network television or cable. This is in part because web-based content is free of the demand to generate massive viewership in any one time or place. There is no prime time. There are very popular shows but no audience share. As a result, web-enabled media doesn't have to be a mass event competing for viewers at any one time. Viewers can watch what they want when it suits them. Viewers can also tune in from wherever they want to—on the beach, in a plane—with nothing more than a smartphone and an Internet connection. Viewers can personally adjust almost everything about their experience. This capacity for customization is what makes the Internet markedly different from cable's 500 channels.

At the same time, the Internet has suspended the design rules of mass entertainment. Anyone can make a video about anything they want and post it for mass consumption. The results are often amazing and bizarre. For example, a fifty-five-second video of a baby repeatedly biting his brother's finger has been viewed over 830 million times. Seventeen million people have learned a new technique of separating egg whites from yolks with a water bottle. A Korean man in a small bedroom screaming at the camera for ten minutes has been watched 440,000 times. When you set free billions of video producers with no rules about what they can make, the result is an increasingly complex range of content.

Web content providers like Netflix, Hulu, and YouTube are successful because they allow hundreds of millions of viewers to tailor their entertainment experience. They call their services "on-demand," but they could just as easily be called "mass customization." By leveraging their capacity for complexity and customization, net-based services are in the process of wiping out standardized, mass-market cable TV. Cable companies have recognized the threat posed by the Internet's capacities for customization. Some are trying to maintain profitability by practically forcing subscribers to bundle Internet connections with cable packages. But these are last-ditch efforts at forestalling the future. Despite the 50 million subscribers that still use their services, the disappearance of traditional cable and satellite television over the next decade or so is inevitable. Given the choice, people will always choose customization.

MASS CUSTOMIZATION

Over the past few decades, the importance of "on-demand" mass customization has become clear. In large part, this has been driven by the Internet's spread of flexible, innovative customization across several industries, including entertainment. But the tsunami-like impact of mass customization is not limited just to web-enabled services. Huge corporations that initially seemed unlikely to be impacted by the wave are also trying to cash in.

The world's most valuable brand, Coca-Cola, is a prime example. For more than a hundred years, Coke has been the innovator in developing and managing a mass-market brand. Beginning in the early twentieth century, the company was the first to demand nationwide standardization of the font and color scheme on their bottles. In 1915, it was the first to design and mandate the use of identical bottles from coast to coast. Following WWII, Coca-Cola became the first truly international brand, selling the same product from São Paulo to Tokyo to Munich. But today even the mass-market company par excellence has been trying its hand at mass customization. In 2014, the company offered personalized messages, including the consumer's name, in the middle of its iconic label. Utilizing new HP printing technology that enables customization at an unprecedented scale, Coke created 800 million personalized labels. These labels were printed with the most common 150 names in each of more than thirty-two European countries, in fifteen languages and five different alphabets. In Israel, Coke took this idea to an even higher level. Using an algorithm created by HP R&D, Coca-Cola printed out two million colorful individual designs. Every single can was one of a kind. This customization program was a breakthrough success, helping Coca-Cola achieve the largest jump in sales since it introduced the twenty-ounce bottle over two decades ago.

Massive supermarket chains like Safeway and Kroger are also gambling on mass customization. But instead of offering uniquely tailored products, they offer individual prices on national brands.

Using data gathered on a shopper's purchases, Safeway sends a list of customized prices via a mobile app. Instead of bulk mailings of coupons, consumers are offered lower prices only on products they buy regularly.

Mass-customized services are also flexing their muscle. The rush to be the "next Uber" has led to apps like Handy for on-demand household jobs, Sincerely to handwrite "personal" letters, and SpoonRocket to deliver high-quality, made-to-order meals faster than the pizza guy. Yes, plenty of start-ups go belly up. Generally speaking, though, as the Internet has made customizable products available to hundreds of millions of people, they have bulldozed standardized mass-market offerings.

3D printed knee replacements, the Internet, and personalized Coke bottles don't have an obvious connection. But they are all versions of the same story. The story begins when a new technology is introduced in mass production. As a result, millions of people got mass-produced knee replacements, cable television, and bottled Coca-Cola, all of which have their own benefits. But, eventually, the same thing happens to each mass product. A second technology comes along that offers the same benefits while adding customization. Knees are still being replaced, but they are biofitted for our bodies. Entertainment is still being delivered, but it is personalized to our tastes. Coke is still being drunk, but out of cans as unique as the individuals drinking them. The Second Wave of a technology offers mass customization. These stories vary in length, but they always have the same ending: Mass customization overtakes mass production.

Today, the exact same thing is happening in manufacturing. Industrial 3D printing is the Second Wave of production technology. It is just beginning to disrupt a $14 trillion global manufacturing industry that is based on mass production. But this is merely the first chapter in an epic novel, a story we've seen play out time and time again. 3D printing is the catalyst for the inevitable Great Disruption of mass production. It's not a prediction, it's a pattern.

A NEW LENS

A 3D printed world is very different from the one in which we now live. It is a place where bricks organize themselves, batteries are the size of a grain of sand, and titanium balls bounce thirty feet into the air. It will result in other miracles that we can't even predict. But though we don't know how long the story will last, we do know how it ends.

We'll start this journey of discovery together by examining why, out of all the advanced manufacturing techniques that exist today, 3D printing is uniquely positioned to spark the Great Disruption. Then we'll look beyond the hype at what the technology is actually capable of today—and what it will do tomorrow. We'll add granularity to the 3D printed world, a place with exotic new materials a hundred times as strong as steel and bridges that print themselves over a canal without any supports.

This is all very cool stuff. But 3D printing's adoption as a common form of industrial manufacturing will be much more profound. The environment, for example, will benefit from a 50 percent reduction in manufacturing waste. Developing countries will reduce their nations' import dependency by developing local manufacturing for the first time. Medical care—from pills to implants—will be customized to every patient's individual biology. Workflow will become more collaborative. Design limitations will disappear.

The point of all this is to answer the two simple questions that lie at the heart of what we would all like to accomplish as leaders, researchers, manufacturers, consumers, parents, educators, business people, and policy makers. How will this 3D printed world unfold in front of us and transform our lives? And what can we begin to do right now to position ourselves and our organizations to our greatest advantage for this incredible future?

2

THE PLAY-DOH EPIPHANY:
A NEW WAY OF MAKING THINGS

The real voyage of discovery consists not in seeking
new landscapes, but in having new eyes.

—MARCEL PROUST

SCOTT CRUMP'S 3D EUREKA MOMENT couldn't have come under more mundane circumstances. He was in his kitchen in suburban Minneapolis, watching his daughter sit on the floor with a big pile of Play-Doh. She rolled it around in her hands, experimenting with different shapes. Then she fashioned a small funnel from a piece of plastic and began pushing the Play-Doh through, like a baker icing a cake. First came one layer, then another on top of it. Eventually, his daughter had built up an entire miniature house structure, leaving spaces in between some of the rows to suggest windows and doors.

Kid stuff, right? Think of Lincoln Logs, popsicle-stick structures, or Legos for that matter. But Scott was an engineer-turned-entrepreneur—open to both process and opportunity—and his insight that afternoon in 1987 would become part of a game-changing technology: His daughter was manufacturing an object not by cutting it out of something larger, like the way that a 2×4 is cut out of a tree trunk, but by layering it from the ground up.

What would happen, Scott wondered, if he did the same thing—but much more methodically and at a much finer scale? What manufacturing possibilities might that open up?

To find an answer, Scott and his wife, Lisa, imagined making a squatting frog by mentally slicing it into horizontal layers, like a stack of poker chips. With a glue gun, Scott squirted out the first layer—the bottom of the frog's feet. After that had dried, he added layer two: more of the frog's feet plus the very lowest slice of its body. And so it went, horizontal layer after horizontal layer, until a 3D frog made out of layered glue sat on the kitchen table in front of them.

Satisfied with his first foray into freehand layering, Scott put his glue gun into an overhead gantry that allowed him to build more precisely. He could move the "printer" back and forth, raising it up into the third dimension as needed. He substituted a fast-drying liquid-polymer resin for glue. Newly available computer design programs added digital precision. With a computer in control, each new layer could be applied without any freehand sketching. Before long, Scott was building all manner of things from the ground up, layer upon layer upon layer.

The technology was disarmingly simple. Putting a glue gun

in a fancy automated machine doesn't sound very innovative. In fact, the technology itself *wasn't* exactly mind-boggling. But from a conceptual standpoint, the introduction of a third dimension to printing changed everything. Take the page you are looking at right now—if you are reading this book in paper. It was created by a 2D printer repeatedly moving back and forth vertically and horizontally. The machine shoots out ink exactly where letters and images are supposed to go. The rest of the space is left white. Your desktop printer probably does the exact same thing.

Now add a third dimension, a *z* axis, to this everyday process. From a technological standpoint, all this requires is a print nozzle that can move up and down and a liquid plastic ink. Again, this technology might not even win the high school science fair. But the printer's capacity is suddenly remarkably different. If you kept printing layers of the letter "o" in the same place with an ink 2D printer, you would end up with . . . an "o." With a 3D printer, the letter would stack up, forming a tube. If a 3D machine printed each "o" in slightly larger font than the one beneath it—ooooo—you would get a small bowl. Use different font sizes and styles every time you enlarge the bowl, and you'll embed a unique pattern on the side of the bowl. A relatively simple addition to the printer has turned you from a writer to a potter and engineer.

Now imagine for a moment that every word in this entire book had been printed on a single page, using plastic instead of ink. Thousands of layers of words would stack up on the page in a dazzlingly complex series of nooks and crannies. By adding a third dimension to the process, you would no longer be printing a book but an intricate sculpture. From the minds of children and an inventor's garage in suburban Minneapolis came an absolutely novel approach to making things.

ADDING IT UP

In 1989, Scott Crump filed a patent for the first Fused Deposition Modeling (FDM) machine, as he called his technology, and started a company called Stratasys. He was not the only operator in the nascent world of 3D printing, though. 3D Systems, a company out of California, was marketing a process that also built in layers. Unlike Crump's extrusion machine, 3D Systems' "SLA" printer constructed objects with an ultraviolet laser and a vat of curable liquid-plastic resin.

Another kind of 3D technology called "SLS" used lasers to melt powders together. In addition to working with plastic resin, these lasers were capable of printing using metallic, ceramic, or glass powders—or combinations of all three. Think of millions of ultrathin welding operations working together to produce a complete object, layer by layer. General Electric, for example, is building those 3D printed jet nozzles out of cobalt-chromium powder. The part is built in layers twenty micrometers thick at a time. That's about .0008 inches per layer, or roughly one-hundredth the thickness of a human hair.

Since the 1980s, many other variations on 3D printing technology have developed. Printers have gotten faster, more precise, cheaper, and use a wider array of materials. But, whatever the degree of technological complexity, all the machines have two basic things in common. They are creating three-dimensional objects, one layer at a time. And they represent a conceptual shift from previous manufacturing processes. 3D printing works from the bottom up, imagining everything as a stack of incredibly thin slices. It is a complete break with how we have made things throughout all of human history.

"**UNTIL TWENTY-FIVE YEARS** ago, there were just two ways to generate a three-dimensional object," says Jeff Hanson, one of Scott Crump's earliest hires at Stratasys. "The first one started with banging rocks together and whittling sticks."

Think about early humans making knives by striking the edges of rock to flake off a sharp edge. Ever since then, we've continued to make things through removing the material we didn't want. We hollow out logs to make canoes. We polish imperfections off of diamonds. We cut off the gristle on choice cuts of meat. We use computer-controlled robotics and sophisticated machines that start, for example, with a block of aluminum, and then cut away all but the finished part. But it's all the same basic process. For millennia, humans have made things by starting with a raw material and then getting rid of parts of it. Today all the processes in which you build something by taking away are known as "subtractive manufacturing." That one category includes almost every way we know how to make things. As Hanson puts it, "We've spent twenty-five thousand years leveraging subtractive processes to generate a 3D part."

The second process is a bit more recent, about forty-five hundred years old. This technique started during the Bronze Age when people began heating up metals until they became liquid. The molten ore was poured into a mold. The process is called "formative manufacturing" because, as it cooled, the ore formed a shape. Swords, rings, and cups were among the first objects made this way.

"Today that has evolved into processes such as injection molding, thermal forming, rim molding, and urethane casting," says

Hanson, "but they are all a formative process and a subtractive process."

3D printing's historic significance is that it is neither formative nor subtractive. Unlike all manufacturing to this point, 3D printing doesn't hollow out or lathe or polish or cast. It doesn't cut or pound or reduce or sand. Instead it creates whole objects from the ground up, layer by layer. It doesn't remove or waste material. 3D printing adds the material it needs to build objects. For that reason, the technology is also known to industrial engineers as "additive manufacturing." Instead of taking away, you build up. It sounds simple, but it is an historic moment in manufacturing. In the past quarter century, humans have come up with a completely new way to make everything. And it opens up an enormous new array of design options.

COMPLEXITY AND FREEDOM

It's impossible to catalog all of the new design possibilities that 3D printing offers to manufacturing. But one way to understand this new design freedom is by understanding the limitations imposed by the old techniques of building by subtraction. From the beginning, engineers have to rule out designing objects with extremely deep cavities because it is practically impossible to get a cutter into them. Trying to mold a design comprised entirely of right angles and straight sides is also inadvisable because it can lead to dried plastics sticking to the sides of a mold. As a result, engineers are encouraged to design objects with rounded corners. There are innumerable similar constraints. As a result, engineers have a whole list of things they *can't do* before they even start a design. With 3D printing, there are no such constraints. The printer doesn't care how complex the object you are trying to create is. It is simply printing things in layers. Suddenly, a whole world of complex geometries can be manufactured. This new level of design complexity means we can solve many problems for the first time.

Take marine biologists' recent success in creating artificial coral reefs. These places are home to an enormous biodiversity, about 25 percent of all fish and other marine animals. But they are also rapidly disappearing as a result of ocean acidification. Over the past few decades, marine biologists have attempted to replace the natural reefs with artificial concrete ones, but the efforts have been only marginally successful.

Part of the problem is that natural reefs have a wide variety of knobby holes and internal spaces that provide shelter for different sizes and shapes of animals. These features are very difficult to replicate with traditional methods. Imagine trying to manufacture a miniature cave that opens up into a chamber on the inside. You could make a rounded mold, but once the concrete sets around it you'll have a problem. You now have to remove the mold from the chamber through a cave mouth that is narrower than the interior space. This is sort of like freezing water in a soda bottle and then trying to pull the ice out through the top. There are solutions, like creating a mold that breaks into smaller pieces for removal. But casting a reef with thousands of complex nooks and holes would mean repeating that cumbersome process thousands of times. And, even then, it was impossible to reproduce all of the unique internal structures of the reefs with a precast piece of concrete.

Then, a few years ago, the first 3D printed coral reefs were installed off the coast of Bahrain in the Persian Gulf. These reefs face no such design constraints. They can be printed with an endless variety of complex variegated internal spaces for fish and other creatures to hide. They can effectively replicate the natural variability of coral reef. As a result, they attract a much wider variety of marine life. The complexity of printed reefs encourages a greater natural diversity. And the reefs can be printed from sandstone instead of concrete, creating lower acidity and a more natural look.

Manufacturing objects based on fractal geometries is another kind of complexity freed up by 3D printing. These designs are the repeating algorithms seen in snowflakes, pineapples, some animals' stripes, and other natural phenomenon. They are captivating to look at, but 3D printing is the first industrial technique to conquer their mathematical complexity. Vanina, a Beirut-based jeweler, prints jewelry using fractals based on tree leaves. Fractal artist Michael Hansmeyer 3D printed the entirety of his haunting 11×52 foot baroque sandstone chapel based on computer-generated algorithms.

Another kind of 3D printed geometrical complexity has proven to be one of the most transformational advances in industrial manufacturing to date. Lattices are the overlapping structures found in the Eiffel Tower or, at a microscopic level, in bones and crystals. If you replace a solid wall with a lattice structure, you can significantly reduce weight and raw materials without any loss in strength. Using 3D printing, lattice structures have been embedded in the build material for everything from airplane wings to bridges to bike gears. In each case, they are lighter and use less material without losing any strength. But these transformational internal structures have to be 3D printed. Traditional subtractive manufacturing will never be able to replicate them.

The advent of 3D printing is historic because it is an additive manufacturing process. It enables us to bring unparalleled complexity into our designs, increasing functionality, reducing costs, and creating new solutions. But there's another reason that 3D printing will be so disruptive. This transformation has to do not just with the fundamentals of subtractive technology itself, but the way these manufacturing processes are used today.

GO FORTH AND SIMPLIFY

Our current approach to mass production began in the 1700s, when the Industrial Revolution mechanized manufacturing. It ended with Henry Ford. Though there have been many notable technology and process achievements since the early twentieth century, it was Ford who proved out today's model of mass production. He created unprecedented economies of scale. He reduced the cost of automobiles to unbelievably low levels. He achieved this through relentlessly applying two fundamentals to his factories: standardization and simplification.

One of Ford's greatest accomplishments was to develop auto parts that were totally interchangeable with one another. Without it, nothing else he did would have been possible. Previously, individual components had to be filed down and fiddled with manually until they fit together. Ford wouldn't accept this compromise. "In mass production," he once said, "there are no fitters."

Once he had fully standardized parts, Ford could build the assembly line he imagined. He placed workers at stations with large quantities of identical components that could be quickly attached or installed. His employees were no longer wasting time working imprecise pieces into place. This meant that manufacturing could take place at a predictable and faster pace.

Next Ford borrowed the idea of a moving production line from slaughterhouses. In these industrial-sized butcheries, cows moved from station to station on hooks. At each stop, a different cut of meat was quickly sliced off. At Ford's factories, a new component was added on the vehicle at each station. Ford continued tinkering with his production until his cars were coming off the line at remarkable speeds. He didn't create the assembly line, but Henry Ford was the first to perfect it.

Ford famously joked that the customer could have any color he wants—so long as it was black. But he didn't pick black as the standard color for aesthetic reasons. It was the only color that could dry fast enough to keep up with his assembly lines!

Ford was also obsessed with waste. The amount of wood scrap sawdust on his floors drove him crazy. It was a mess that had no place in a standardized, simplified factory. He grew so frustrated that he began to bundle up all these scraps in briquettes and sell them. He eventually spun this product out as a new company, and named it after his brother-in-law, E. G. Kingsford.

As he squeezed new efficiencies out of simplification and standardization, Ford's production lines kept getting faster. By 1915, the time it took to build an automobile had decreased by 90 percent. As the number of cars increased, the costs dropped proportionately. When it was introduced in 1908, the Model T cost $825. By 1912 it was down to $575 and by the mid-1920s the price had fallen to $290. The low prices changed the makeup of American roads. In 1916, more than half of the automobiles in the country were the exact same model: black Model Ts.

Like many other industrialists of the day, Ford saw his goal as nothing less than extending democracy to commerce. And his techniques were remarkably successful. Even Ford's blue-collar workers could now afford a product that, fewer than two decades earlier, had been a plaything of the rich. Ford wasn't just gaining market share; he was creating a whole new market. He mastered assembly-line production and, as a result, Ford created the world's first mass-consumer product.

With his resolute determination to pursue scale, standardization, and simplification, Ford punctuated the Industrial Revolution. Since the early twentieth century, his principals of mass production have expanded to define every part of the world around us. All innovation has taken place in the margins of Ford's accomplishment. In the automobile industry, for example, companies like Toyota and GM have made their own improvements over the past century. But they still rely on massive assembly lines to create mass products. Apple, one of the most valuable, innovative companies in the world, makes products Henry Ford couldn't have dreamed of. But he could have predicted exactly *how* they would be made. The truth is that, even with the addition of robotics and advanced materials and computer modeling, manufacturing hasn't experienced a global paradigm shift since Ford ushered in the world of modern production.

The problem now is that Ford's brilliant plan has been exhausted. The largest manufacturing companies still constantly look to squeeze efficiencies by simplifying and standardizing production. With all this time, money, and intelligence invested in improving production, it may come as a shock that modern manufacturing is still a very inefficient global system. And fixing these problems doesn't require squeezing harder. We have pushed this equation of scale, simplification, and standardization to its extreme limits. We've run Henry Ford's mass production into the ground. But we're not completely out of options. What if Ford's model is not the final word on manufacturing? What if a technology came along that could flip everything upside down all over again?

BEYOND SCALE

Here's where things get interesting. Large-scale manufacturing created the mass market by constantly increasing scale and standardization—a million black Model Ts. Its advantage is predicated on producing ever-larger numbers of products. But with 3D printing, costs are not tied to the number of units produced, so you are not penalized for producing in small quantities. The cost per-unit line for 3D printing is flat.

The initial cost of producing things with 3D printing was high, particularly when compared to the low unit prices achieved with efficient mass production. Today, 3D printing is still only cost effective for a small range of industrial manufacturing. The initial quality of 3D printed objects was also low. But the same could be said at the introduction of nearly every other transformative technology. And it is rapidly improving.

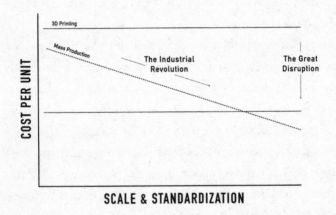

SCALE & STANDARDIZATION

From 1990 to 2000, the quality of the printers increased, the costs decreased, and, for the first time, the industrial application of 3D printing became feasible. Since then, 3D printing has been used primarily for rapid prototyping, a process that allows designers and engineers to quickly create a physical representation of the object they are designing. The market for 3D printing used for rapid prototyping has grown quickly to several billion dollars in the

United States alone. Stratasys, Scott Crump's company, is the leading 3D printer manufacturer with a market value in the billions.

Today, industrial 3D printers are producing parts with precisions and tolerances comparable to traditional manufacturing. Behind closed doors, an increasing number of companies have begun aggressively experimenting with this technology in areas well beyond rapid prototyping, such as on-demand production to replace physical inventory, rapid design iteration, and customization.

As this trend of falling costs and rising quality accelerates, there is only one possible outcome: More and more of the conventional manufacturing value chain will become vulnerable to the superior design flexibility and customization available only with 3D printing. At the same cost and quality, designers and engineers will always prefer the full range of complexity of 3D printing manufacturing that is available to them. Consumers naturally prefer customized products that fit, work, look, or function exactly as wanted, and are produced when, where, and in the exact quantities desired. As price continues to decline, 3D printing will displace greater and greater areas of conventional manufacturing. This is what changes everything. Ultimately, the rise of additive manufacturing will leave the Industrial Revolution in the rearview mirror. The Great Disruption has begun.

FORD REINVENTS MANUFACTURING

In 1988, the Ford Motor Company purchased an early SLS 3D printer, the third ever made. The leader of their advanced manufacturing group was fascinated by the new 3D printing technology. He continued to snap up the newest machines as they came out. Next, Ford bought the first LOM or Laminated Object Manufacturing, a machine that built three-dimensional objects using paper laminates. In fact, the company paid for a machine that didn't yet exist, financing the research and development of the LOM.

A company called DTM had also developed an early version of laser printing that could make 3D objects out of metal. The product development was financed with the help of DARPA, the same U.S. military group that played a critical role in what became the Internet. Ford bought one of those metal printing machines as well as Stratasys' FDM model that printed in plastic.

With all of this cutting-edge technology installed, Ford's Advanced Manufacturing Center would invite in company engineers and train them on the operation of a machine for a week. Once they were trained, they could use the center whenever a machine was available. Many returned to their department and requested machines be purchased for on-site use. The problem was that, although the company owned an impressive amount of hardware and the technology had piqued the interest of some engineers, they weren't really being used. It took the persistent curiosity of a few Ford engineers to move the company's use of 3D machines to the next level.

Harold Sears is a twenty-five-year veteran of Ford who originally joined the company as an expert in Computer-Aided Design. CAD—as it is known—is a suite of programs that assist engineers in designing 3D objects without ever physically touching them. This technology sounds run-of-the-mill in a world where novices can edit music, photos, and movies. But in the 1980s, computer design programs were just replacing drafting tables and pencils at many large manufacturers. Harold's division at Ford had also bought a 3D printing machine, trained people on it, and then locked it up in a closet. About six months later, Sears approached his boss and expressed interest in running the machine. His boss, who Sears described as "a real techie," was excited.

"Good," he said. "I can't stand this thing being locked up."

Sears began to help engineers quickly convert their computer-aided designs into digital files that the 3D printers could understand. Soon Sears moved out of his computer programming position and began operating a 3D printer. Today he is a technical expert for additive manufacturing technologies. Over that same time, Henry Ford's company became a global leader in embracing and innovating with these technologies.

Originally, Ford's engineers looked at 3D printers much like Scott Crump did: machines that could be used to create models very quickly. "It was just called rapid prototyping because that's all they really thought about it," said Sears. But in the past four or five years, the technology has made a significant leap forward, becoming more professional, with higher quality and higher volume. The new machines still print in layers, but the end product is more advanced than that used for rapid prototyping. The more mature technology isn't limited to mere design models. It can be used to make testable models with strengths close to end-use parts. It can even be used to make final products.

Today Ford has five additive manufacturing centers churning out an enormous number of units. Sears' office sits in a facility that turns out twenty thousand 3D printed parts annually. While about 80 percent of these components are still for prototyping, there is a growing percentage used in the rest of the production chain. This is a huge change in how things are designed and built.

The traditional design process that Ford uses is very similar throughout manufacturing. First, a team of engineers design, a product and orders prototypes for it. Then the prototype goes through a series of tests for safety, strength, styling, performance, et cetera. Once the part is approved, Ford orders all the unique components necessary for mass-producing huge volumes of the part. This is an extremely expensive proposition. No matter what it is, any newly designed part will require specific equipment and special tools. These might include molds for casting foam or punches, cutters, and rollers for shaping metal. Whatever the specific tooling, it is all uniquely designed to the specifications of the approved prototype.

This is where manufacturers make their biggest gamble. If the part works out like the engineers hoped and the company sells tens of thousands of them, they win. But if there is a problem with the part and they need to go back to the drawing board, manufacturers could easily be stuck with hundreds of thousands of dollars of useless tooling. These problems are far from rare. More than half of parts go back for some sort of design changes. Even if it guesses correctly, the company is stuck using this same design for years, even decades, while competitors introduce newer, better designs into the market. By betting long on scale, companies tie up capital, take on huge risk, and often end up with huge losses.

Additive manufacturing skips all these time-consuming and expensive steps. If an engineer decides that there is a flaw in the design, you simply open up the original computer files and make adjustments until you print out the right one. If the market does not react to your design in the way that you expected, change it! This flexibility goes a long way toward explaining how a new technology is already being used to produce the end-use parts on some of Ford's iconic vehicles, like the F-150 and Mustangs.

"This is very exciting for us," says Sears. "When you build a prototype part, you save money on the cost of the part, and you usually turn it around quicker than it could have been done conventionally. That's end-of-year savings."

3D printing, a brand-new way of making things, has gained a foothold at Ford and innumerable other companies. Ford uses 3D printing to produce nearly one hundred thousand prototype parts per year. Harold Sears and other engineers use the technology because it can be faster, cheaper, or create complex geometries that no other manufacturing process can. It helps workers do their jobs more quickly and efficiently. In short, it is a useful complement to traditional manufacturing techniques. For years to come, 3D printing will coexist with traditional manufacturing techniques as an incredible new tool with more and more viable applications. But, eventually and inevitably, it will begin to overtake them.

IT'S 1994 AGAIN

None of this 3D printing technology is a secret. But most of us don't really know what to make of it. The media tends to over-hype consumer-grade printers, without going into the true capabilities and potential for higher-end 3D industrial printing. As a result, most still don't understand the "what" and "why" of 3D printing. It's not just a hobbyist making a cool gift for a kid. It's more than marine biologists trying to save coral reefs. It's a technology that's just a few decades old but is already displacing traditional, subtractive, mass-produced ways of making things. Ultimately, the spread of 3D printing is about the disruption of the manufacturing processes that humans have been perfecting for centuries.

In a way, our obliviousness to this historic event is reminiscent of the earlier days of the Internet era. In the early 2000s, I ran a CEO networking company. Around 2003, I remember that my clients knew that an "Internet Revolution" was changing business practices. But they still hadn't really grasped the scope of this transformation. A lot of them were still talking about how banner ads were a primary implication of the web. These were very smart people. But nothing had made them fully aware that a disruptive phenomenon was already remaking the world around them.

A large part of this disconnect was because they hadn't yet engaged with the technology in a meaningful way. At the time, everyone had Blackberries. Virtually no CEO carried around an actual laptop, so they didn't necessarily use the Internet for anything more than emailing. Then the iPad came out and these executives started carrying them everywhere. Unlike a laptop, which gave the impression that you weren't above the work, the iPad was a status symbol. It represented a tool that allowed a leader to survey what was happening, while still delegating everything. This was when it all clicked. As soon as they had the tools of technology in their hands, their perspective completely changed. They had a much better understanding about what the Internet could mean and where it could go.

I see the same thing happening with 3D printing in the next few years. The basic technology exists. It's been around for a few decades—long enough that when I talk to the engineers, they think that everyone knows about it. But the nontechnical senior-level people I talk to are just waking up to 3D printing's possibilities. When it clicks, it's like Christmas morning for them. They are fascinated by how many of 3D printing's applications have far-reaching implications. They are eager to invest in equipment and start pilot programs. They are assigning people and teams to start exploring 3D printing in a meaningful way. What they see is how the technology's complexity and customization—the same characteristics that are already transforming medicine—can disrupt the rest of manufacturing.

The Internet's early promise was "information will be free." It was simultaneously so vast and simple that it was impossible to fully comprehend. Sure, you could communicate more frequently and easily. An email was cheaper and much faster than mailing a letter. You could find information almost instantly. Some people were even beginning to realize the Internet's enormous retail possibilities. But in the early days, no one imagined anything like a shared economy in which Uber and Airbnb would disrupt the taxi and hotel industries. No one expected the huge amount of micro-funding enabled by crowd sourcing, with companies like Kickstarter empowering an entire new generation of entrepreneurs. No one predicted companies like Twitter and Facebook would empower thousands of young Egyptians to overthrow their well-entrenched government.

The same is true of the deceptive simplicity of 3D printing's promise "you can make anything in layers." As this manufacturing process spreads and matures, it is easy to guess that companies that make, sell, and deliver things will adapt to leverage the technology. But what do we make of a world in which nearly every product on earth can be uploaded and manufactured locally, or even on-site? What are the implications of machines that can print whole houses in wood? When tablets that include metal and plastic, wiring and circuit boards, battery and transistors can be printed all as a single object? When pipes can be printed that automatically repair themselves when damaged? When food can be printed in mass quantities and unlimited tastes and textures in the farthest corners of the earth, or in space? When a plane can print another plane in the air and drop it in flight?

These examples seem ripped from science fiction, but every

one of these concepts is real and currently in development or testing. More important, these examples don't even begin to explain what 3D printing will eventually mean to the world. Just like the disruptive changes wrought by the Internet, understanding and anticipating the impact of such a momentous shift is no easy task. Predicting the exact events of the future is impossible. But, by looking back at previous disruptions, we can begin to envision with surprising clarity where this technology is taking us.

But don't take my word for it. This is not a prediction. It is a pattern. In fact, it is the same pattern that we see with the introduction of every major technology of the last two thousand years. To help add some color to our 3D printed future, let's first take a step back and look at the some of the greatest technological leaps in history.

3

THE SECOND WAVE

There are patterns which emerge, circling and returning
anew, an endless variation of a theme.

—JACQUELINE CAREY

IN 1438, A WOULD-BE ENTREPRENEUR named Johannes was walking around Aachen, in present-day Germany, with an extra spring
in his step. The church leaders in the town had just announced
plans to host a massive religious celebration. In exactly one year,
tens of thousands of pilgrims would descend upon Aachen to
gaze upon the cathedral's four great relics: Christ's swaddling
clothes, his loincloth, Mary's cloak, and John the Baptist's behead-
ing cloth.

This was an enormous gathering by the standards of mid-fifteenth-century Europe. For Johannes, it represented an irresistible market opportunity. He devised a scheme to make and sell tiny mirrors to the pilgrims, which he would market as having wonderful religious powers, a "holy glow" that could be carried back home to sick family members unable to make the journey. Johannes convinced three investors to bankroll his new venture and with their help purchased large quantities of lead, tin, and antimony as raw materials. With everything in place, Johannes set out to make the holy vessels.

But before he had completed even his first mirror, an outbreak of the plague fell upon Western Europe, the leaders of Aachen decided to cancel the event, and his investors showed up on his doorstep demanding their money back. His back against the wall, Johannes knew he needed a new plan and a better idea. Then one struck him. A skilled metalworker, Johannes proposed to his investors that he use the raw materials purchased for mirrors toward a completely different purpose. Amazingly, his investors agreed and decided to double their investment. This time he delivered in spades.

One year later, Johannes Gutenberg had created his first printing press with movable type. His invention, known as the Gutenberg press, is consistently ranked among the top five most important inventions of the last millennium. The technology churned out books and other printed matter at an unprecedented rate. Within just a few years, it began displacing traditional craft production of books, and his investors were thrilled. But the invention proved to be much more than a successful business venture. Within a few decades, the printing press also created political upheaval and social change on a massive scale. The world would never be the same.

Printing presses existed for centuries before Gutenberg's birth. But his inspired tinkering created machines that were so much faster that they changed the course of history. Gutenberg's primary breakthrough was dividing up printing into standardized component parts. He created different inkable stamps for each letter and punctuation mark, as well as well-known abbreviations. All this sounds like common sense today, but mechanical printing in the mid-1400s meant laboriously carving whole wooden blocks for each page. Gutenberg's small casted glyphs were much more nimble. They could be rearranged endlessly to create any word, sentence, and paragraph. Instead of carving one-off stamps for a whole page, printers could easily lay out a page using copies of a few dozen interchangeable letters and punctuation marks.

Not only was Gutenberg's new process far faster, it was also much more efficient. In traditional printing, the page-sized wooden stamps were discarded once the print run was finished. Printers using the new interchangeable type simply rearranged the exact same glyphs on the next page.

Gutenberg's next step was to massively expand the scale of printing. The commonly used wooden plates wore down over time. They weren't built to make more than a few hundred copies at most. Gutenberg's new metal type could withstand thousands of reprintings. The inventor also developed higher-quality inks that didn't smudge. He sped up production time by adapting the screw presses traditionally used to crush grapes for wine. When a whole page of text was arranged and inked, the printers pulled a lever that forced a heavy plate down on top of the paper. The process was simple, eminently scalable, and required a minimum of manual labor.

Gutenberg used standardization and scale to create a radically more efficient printing process. The result was an unprecedented flood of literature that dramatically lowered the costs per unit. This boom created the conditions for the first mass market for books. The new industries of mass production, distribution, and sale of books soon developed—the antecedents of today's publishing houses, booksellers, and media empires.

BEYOND WORDS

But this sudden expansion of access to written knowledge also had huge social and political implications. Before Gutenberg, the Catholic Church had closely guarded its nearly complete control of literature and written knowledge. As the new printing presses spread across Europe, Bibles, political pamphlets, treatises on philosophy, and Latin grammars became cheap enough that members of the fledgling bourgeoisie could afford them. Europe was just beginning to emerge from the isolation and provincialism of the Middle Ages. The press sped up this process tremendously. Suddenly, ideas about astronomy, agriculture, medicine, government, art, and religion that had been tightly controlled—or essentially lost for centuries—could be shared widely across the continent in months. The explosion of available information powered the Enlightenment, the Scientific Revolution, the Age of Discovery, colonialism, and, ultimately, the modern world.

There was one other great historical impact of Gutenberg's press that he could not have imagined. By all accounts, Gutenberg was a faithful Catholic. He saw his more efficient, standardized press as a way to spread the church's word on a grand scale. His lasting masterpiece is the Gutenberg Bible. But his printing press was such a transformative technology that it spurred revolutions, which, by their nature, are hard to predict and control.

By the early 1500s, Gutenberg's technology was being used to print huge runs of Martin Luther's 95 Theses, one of the very first public condemnations of the church. Within months, people from Florence to Paris to London had read the priest's polite but stern criticisms of church practice. By then, printing presses had been established throughout the continent. The document's message spread across Europe faster than ever previously possible. By the standards of the 1500s, it went viral, ultimately sparking the Protestant Reformation. By the end of the century, Gutenberg's press was largely responsible for the church's loss of England and Scandinavia as well as large parts of Holland, Switzerland, and Germany.

FROM HOT TYPE TO NO TYPE

The movable type printing press went on to dominate the production of literature and newspapers for centuries, but it was not the last transformative information technology. In the 1960s, the United States Defense Department commissioned development of a new, robust communication system designed to link military and research computers. A few decades later, masses of people around the world began using the Internet to get their news or download books. Between the mid-1990s and the early 2000s, the Internet completely disrupted the newspaper and publishing industries that had been spawned by Gutenberg's press and remained dominant for the next five hundred years.

The printing press and Internet both had world-changing impacts on the creation and distribution of information. But the Internet was not just a supercharged version of the printing press. In fact, the technologies impacted the world in very different ways. Why is that?

Gutenberg's invention was a revolution, a phenomenon with

specific characteristics. First, it democratized information. For centuries, access to written knowledge had been held by a tiny percentage of the population, the church and its universities. In 1450, the total number of printed books in all of Europe could fit on the shelves of an airport bookstore. Following Gutenberg, a huge amount of literature became available to a much wider audience. This same democratizing dynamic exists in any revolution. An existing utility—from written information to agricultural innovation to political power—spreads from the few to the many. The printing press was revolutionary because it wrested a long-existing utility, written literature, away from the control of a small group.

Revolutions all achieve their democratization of a utility in the same way. They replace small, individual craft production with scale and standardization. Most books in Gutenberg's time were still made by individual craftsmen. Each book was a unique product. The resulting work, frequently embellished with elaborate illustrations, was often magnificent. But even after centuries of innovation in paper and inks and writing utensils, craft production was still laborious and slow. In the margins of some bibles from the Middle Ages, copyists left agonizing notes: "I am very cold," "Oh, my hand," and "Now I've written the whole thing: For Christ's sake give me a drink!" Gutenberg's invention got rid of one-at-a-time production. On a printing press, every book in a five-hundred-volume run is exactly the same.

As the press created a mass market for cheaper books, some craft producers rebelled. Members of woodblock printing guilds destroyed presses and tried to have them banned. But the craftsmen could not compete with the power of scale and standardization that Gutenberg's press unleashed. As the new technology's scale forced price per units ever downward, it was quickly adopted across the continent. In 1450, there was one Gutenberg press in Europe. By 1500 there were roughly one thousand, from Krakow to Rome to London. The fall of craft printing was inevitable.

Achieving massive scale also adds an initial hurdle to the reproduction of information. It requires more start-up capital than craft production. A scribe with paper and pen could produce just a few copies of a book much more cost effectively than a press. Investing in the capital-heavy new technology just to print limited runs was a sure-fire money loser. But that's not what the press was designed to do. Its purpose was to create large numbers of identical volumes, while massively reducing their per-unit costs. And it was remarkably successful. In the fifty years following the innovation of Gutenberg's press, the cost of books dropped by 90 percent. By 1500 there were 20 million books in circulation, or one for every five Europeans. This brings us to the final characteristic of a revolutionary technology. After centuries of high-cost, low-volume craft production, revolutionary low-cost, high-volume production drove market penetration unlike anything previously experienced.

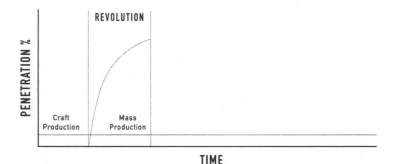

Technology "revolutions" bring the utility of a new technology to the masses for the first time.

THE INFORMATION DISRUPTION

Though the technology has also been transformative the Internet is decidedly not a revolution. It is a disruption, a phenomenon with qualitatively different dynamics from revolutions like Gutenberg's press. For example, the printing press created an abundance of information—but it was a one-way street. Access to the infrastructure and production of this information was still tightly guarded. The press was democratizing in one sense, but the vast majority of people still had virtually no chance of getting a book contract or becoming a newspaper columnist. The likelihood of owning a mass-market newspaper was even lower.

This all changed with the Internet. Over just a few decades, the technology disrupted this hierarchical model of information creation and sharing. The Internet virtually eliminated gatekeeping. It makes the masses of people not just consumers of media, but creators. Today, hundreds of millions of people are blogging or self-publishing. Readers have become writers. Their literature is readily accessible to a billion people. Disruptions expand direct access to the technology itself.

Second, disruptions expand direct access by lowering the entry barriers. It is so cheap—frequently free—to post content online that virtually anyone can do it. Previously, the start-up expenses associated with scaled production of books or newspapers made this kind of direct access virtually impossible. The Internet's comparatively low-entry barriers drive today's profusion of self-produced books, movies, advice columns, and music.

Finally, disruptions result in a movement from simplicity of design toward increased customization and complexity. The economics of mass production—sell as much as possible of the same thing—mean most publishers and movie houses want products with popular appeal. But another way of describing these products is "generic." Mass-produced content strives for easily understandable simplicity. This results in the constant use of well-worn conventions. In romantic comedy novels, the primary characters meet, are separated, and then, through some trial, are reunited. The same is true of genre movies. When Roald Dahl wrote the screenplay for the James Bond film *You Only Live Twice,* he was given relatively free rein, except two strict rules. First, there had to be three "Bond girls"—an ally, a henchwoman, and a main love interest. Second, the first two had to be killed over the course of the movie.

The people creating and selling this mainstream material are very capable of more complex stories. They are simply aware that their industry is based on scale and standardization. Big media and entertainment companies are only profitable when they sell huge amounts of the same thing, be it copies of James Patterson books, episodes of *Law & Order,* or franchise movies like *Star Wars.*

Internet programming is not tied to any such rules. YouTube videos that consist solely of a screen view of a video game narrated by amateur players joking around attract millions of viewers. Equally huge numbers have watched clips of films reedited with different dialogue or audio. These bizarre parodies could never exist in a world oriented around mass-market production economics. Disruptions always encourage radical customization.

What is interesting about the rapid spread of quirky, personalized content is how popular it has become. Take the over three hundred hours of video that are uploaded to YouTube every minute. The vast majority of this material is esoteric and amateur. It will never even be seen by more than a few dozen people. But some of it will be seen by hundreds of millions of people. This customized content is particularly popular among younger people who have grown up in the Internet era. According to *Variety* magazine, the five most influential figures for American teenagers are all self-produced web-based personalities. These amateur YouTube celebrities beat out mass-market singers and movie stars like Katy Perry, Jennifer Lawrence, and Seth Rogen. The most popular is Smosh, two twenty-eight-year-olds who met in sixth-grade science class and started doing web shows for their friends.

Comparing the most popular print magazines with the most popular websites is another measure of how the Internet drives radical customization. General interest titles like *People, Time,* and *Better Homes and Gardens* are typical of the highest circulation magazines. While the most popular online destinations include a few general interest and celebrity gossip sites, they tend to be much more esoteric. Take these headlines from the splash pages of a few sites: "Bitfusion's New Cloud Adaptor Lets Developers Use GPUs and FPGAs in the Cloud," "My Blinking Light Will Change the World!" and *"Dying Light* Creators Raise Price of Expansion in the Classiest Way Possible." There is a pretty good chance that you have no understanding of, or at least have no interest in, one of those headlines. But the sites all attract over ten million unique viewers a month.

Post-Internet content has become increasingly customized and complex. The result is a media landscape that is more challenging for traditional scaled-content industries. Publishing houses have to compete with millions of self-published titles every year. Television networks have reduced their expectations of total market share. As audiences continue to choose the radical customization of the Internet, the information and entertainment industries will face increasingly difficult challenges. Among younger consumers, the Internet is already beginning to make the top-down distribution of entertainment irrelevant. Eventually, nearly all content will be created by dispersed sources. At the same price, people consistently choose the advantages of complexity and customization. Disruptions inevitably upend revolutions and, with them, their standardized and scaled industries. This cycle of revolution and disruption reoccurs throughout history.

REVOLUTIONS AND THE IRON HORSE

The prototypes of the first steam locomotives were generally considered exciting novelties. There seemed little chance that they could ever displace the horses that pulled plows through fields, carts out of mines, and coaches around the country. Then the cost of owning horses skyrocketed during the Napoleonic Wars. In the early 1800s, some coal mine owners experimented with using the mechanical contraptions to haul their product out of the earth. Even then, the transition was uneven. Because they relied on huge reservoirs of high-pressure steam, locomotives were prone to explosions. Once the technology began expanding, they also drew popular resentment by replacing the beloved horse. Nonetheless, by the middle of nineteenth century, steam-powered trains became the primary means of rapidly transporting goods and people throughout the country. In 1844, an act of Parliament required each train company to provide at least one cheap train per day, at a cost of no more than one penny a mile. Cities, towns, and villages were linked through an intricate web of railway lines stretching right across the country.

Trains are another classic example of a technological revolution. They democratized personal mobility for the first time. In early nineteenth-century Britain, the fastest means of transportation were the Royal Mail coaches. In theory, these carriages were public, but only the wealthy could afford them. The private coaches that galloped along British highways with family crests on their side were even more exclusive. For most people, personal mobility was limited to walking or traveling by horse. An overwhelming majority of the country lived, married, worked, and died within a few miles of their place of birth. Then, within a lifetime, most of Britain was able to travel anywhere in the country. This mass mobility, in turn, accelerated migration to urban centers and factory jobs and thus stoked the Industrial Revolution.

As with the printing press, this revolution in personal mobility was driven by scale and standardization. Obviously, trains always took the same routes on the same tracks. But time was also standardized by the trains. In the first half of the nineteenth century, it was generally agreed that, say, noon was when the sun was highest in the sky. But there was no national authority that tightly linked Cardiff clock time with that in Newcastle. This was not good enough for British train operators. A few minutes of variation every few towns would screw up their schedules. The companies invented standardized timekeeping for the whole country. Initially, towns might have two clocks—one for "local time" and one for "train time." Within a few decades, though, every city in the country switched to train time. The locomotive became the first enforcer of a standardized national time zone. Like the printing press, the revolutionary locomotive expanded access to a utility through standardization.

The train also drove its horse-drawn competitors out of business. Coach companies carrying just a few passengers at a time could not compete with the more efficient, faster, and cheaper steam train. Once again, a scaled technology overwhelmed a personalized industry.

COMPLEXITY AND CUSTOMIZATION

The train was not the final word in transportation, though. In 1919, Henry Ford opened a Model-T production facility two hundred miles north of London in Manchester. Ford's new product, the first affordable automobile, offered the British suburbanites customized personal mobility far beyond what track-bound locomotives could. The introduction of the automobile had enormous impact, rewriting city maps and changing social mores. London's outer neighborhoods swelled. In Los Angeles and other cities around the world, the entirety of the urban experience was developed around the technology. The automobile was a disruption of transportation. It turned riders into drivers, with all the autonomy that implies.

By lowering entry barriers, Henry Ford's product also gave people direct access to the means of transportation. Only a handful of people could afford their own train network. But by 1920, automobiles were affordable to the majority of people in the United States, Great Britain, and other wealthy nations.

Finally, the auto shifted personal mobility from design simplification toward complexity and customization. A train or streetcar follows a relatively simple route over and over again. Drivers use their cars to completely customize their transportation. Even if they primarily use their vehicle to commute to work, they personalize their beginning and end points. And they can easily leave their predetermined path to go grab groceries or stop by a friend's house if they want. This same self-navigation is happening millions of times across the United States every minute, creating a much more complex transportation landscape.

As with the Internet, this customization quickly created challenges for the existing revolutionary rail technology. In 1910, after half a century of feverish expansion, the United States had one of the great train transportation networks in the world. That year, passengers took nearly a billion trips around the country on over 350,000 miles of track. Trains accounted for nearly 100 percent of the passenger traffic in the country. But train travel was nearing its peak. After WWII, the production of automobiles and the highway system expanded. Meanwhile, train ridership began declining significantly. By 1957, only about 30 percent of passenger traffic was on trains. In half a century, automobiles had stolen 70 percent of the trains' people-transportation business! The problem wasn't that the train companies were horribly mismanaged or offered a worse product. They were just fighting a losing battle. At the same price, people strongly prefer customization and complexity. The decline of the passenger train was as inevitable as the Internet's disruption of publishing.

REVOLUTION'S GAMBLE

Disruptive technologies like the automobile and Internet will always overwhelm revolutions for one simple reason: Revolutions live and die by scale and standardization. Their great contribution to a utility—expanding access by lowering prices—is done at the expense of customization. Revolutions democratize a utility, but they eliminate individualized products and services. Printing killed off the production of single volumes. Trains don't take you home. A disruptive technology offers the benefits of craft production, customization, to the mass market created by revolutions. Any technology that offers mass customization of a utility will inevitably disrupt a method based on scale and standardization—no matter how entrenched the industry may seem.

When we chart the changes in market penetration in both

information and transportation, we see what looks like two waves. For centuries, there is a flat line of low-productivity craft production. Then along comes mass production, a revolutionary swell that remakes the world. But the story isn't over. Eventually, a Second Wave—mass customization—appears that upends the world once again. The Second Wave gains market penetration at the cost of the First Wave. But Second Wave technologies like the Internet and automobile aren't the opposite of First Wave technologies like the printing press and train. They are the second act of the same cycle of innovation.

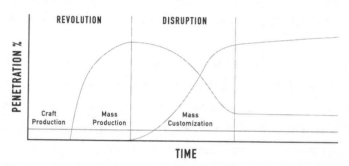

As in every mature revolution, the disruption of the information and transportation revolutions was inevitable. Customization is the natural Second Wave of technology adoption.

This is where 3D printing enters our story. The Industrial Revolution dramatically accelerated production of manufactured items, lowering their per-unit costs. Huge numbers of people gained access to a new range of affordable goods, from stoves to clothing to guns. For the first time, masses of people lived in an era of relative material bounty.

The Industrial Revolution also accomplished these goals through relentless standardization. Henry Ford is the preeminent example of this trend, but it was widely applied throughout industry. A mechanical engineer named Frederick Taylor even developed a science for industrial bricklaying, shoveling, and lifting pig iron. Taylor's teams observed workers at factories. Then they taught workers how to improve efficiency by trading in their own individual motions for one standardized technique.

Over the course of the Industrial Revolution, large segments of the market for craft goods were destroyed. Individual producers of frying pans or blankets couldn't match the efficiencies of scaled and standardized industrially produced goods.

In short, the Industrial Revolution fits every criterion of a revolutionary technology. But it was just the First Wave of mass production. Until now, manufacturing has been waiting for its Second Wave. 3D printing is that technology. It trades scale and standardization for customization and complexity. It lowers entry barriers into manufacturing—desktop 3D printers can be had for $500. As a result, it increases access to the creation of manufactured items. The result is a flood of new designs for everything from customized phone cases to prosthetic arms to printable electronics.

From the automobile and Internet, we know that, at the same price, people will always choose customization and complexity over the design limitations of scale and standardization. In some areas of manufacturing, 3D printing has already achieved price parity and is displacing traditional manufacturing. As the technology matures, it will inevitably disrupt more and more conventional industrial manufacturing—no matter how entrenched the technologies may seem.

THE INCREDIBLE DISAPPEARING SEGWAY

On December 3, 2001, the future of transportation was unveiled live on *Good Morning America*. In the months leading up to the event, there was speculation that the device, codenamed "Ginger" or just "IT," was a personal hovercraft or antigravity machine or some other sci-fi technology. Steve Jobs said it would be as important as the PC. John Doerr, the venture capitalist behind Netscape and Amazon, said it would be more important than the Internet. The inventor, Dean Kaman, had a wide array of impressive inventions to his name, including a groundbreaking syringe and dialysis equipment and an all-terrain wheelchair. But his newest technology, the Segway, ended up being a slightly awkward-looking scooter.

After riding a prototype around Bryant Park, host Diane Sawyer's reaction was, "I'm tempted to say, 'that's it?'"

The problem was not technological sophistication. The Segway combined advanced gyroscopic technology, electronics, software, and mechanics in ways that had never previously existed. It worked as the inventor intended—it was an extension of the human body that adjusted speed and direction with only subtle commands. Why was it such a massive letdown?

The Segway's problem was that it didn't meet the essential criteria for a transformative technology. It wasn't revolutionary—in the sense of expanding access to an existing but closely held utility. Neither was it disruptive. It didn't allow people to customize their travel experience in any profound way. In densely urban areas, people can walk or ride bikes and use subways, buses, and cabs. In less-dense areas, they can drive cars. The Segway was a cool technology, but it didn't really expand, augment, or improve our transportation options. Today the once-hyped invention is mostly used on guided tours, by police, or by stockers in large warehouses. It didn't—and never will—change the world because it doesn't meet the criteria to do so.

3D printing is another sophisticated technology that is often overhyped—or talked up for the wrong reasons. Engineers are quick to point out its mechanical or economic shortcomings. You can't print an entire car, at least not yet. It will certainly not completely replace traditional manufacturing in the next few decades. But, unlike the Segway, we know that the threat posed by 3D printing is truly and uniquely disruptive. It is the only technology capable of moving manufacturing from a model based on standardization and scale into a new era of customization and complexity. It lowers entry barriers, turning users into makers. It throws off design limitations and thrives on complexity and customization. So, while the horizon for many transformations is much further out than is sometimes believed, 3D printing will also eventually disrupt and displace much of traditional manufacturing. It will do so as inevitably as the Internet cut down publishing and Ford stole passengers from Union Pacific.

Industrial 3D printing is manufacturing's Second Wave, the technology that will trigger the Great Disruption, but these are still early days. As it matures over the next few decades, we will see an unprecedented restructuring of the global economy. There's no better proof than seeing what 3D printing can do already.

4

MASS CUSTOMIZATION

**The despotism of custom is everywhere the standing
hindrance to human advancement . . . .**

—JOHN STUART MILL

ON MAY 11, 1880, a dozen armed settlers left a community picnic
to confront representatives of the local railroad. The railroad,
Southern Pacific, had offered the men a deal on land in California's
San Joaquin Valley. The settlers could build improvements—like
wells and buildings—on the land. In exchange, they received as-
surances that they could buy their plots in the near future. The
land was particularly attractive because it abutted a future rail
line. In nineteenth century California, wealth, people, and goods

moved along rails. Like many early migrants to the state, the settlers were prospectors. But instead of gold, they were prospecting for land.

By the day of the picnic, two things had soured the men on the railroad's deal. First, the Southern Pacific demanded higher prices on the land than the settlers expected. Then, the rail company changed the route so it completely circumvented the settlers' territory. Stuck in the middle of the San Joaquin Valley with no rapid way to transport themselves or their crops, their homesteads would be nearly worthless. The men considered themselves victims of corporate whims and broken promises. A court case was pending. Then, during the picnic, the settlers got word that four railway men were going from homestead to homestead demanding the settlers pay Southern Pacific's asking price. If they refused, they were served with immediate eviction notices.

Later testimony suggested that, despite their weapons and anger toward the company, the settlers had not intended a violent escalation of the encounter. Instead, they hoped to convince the railway representatives to hold off on evictions until the court case was settled. But personal animosity between two men flared up. After a heated exchange of words, they drew their guns simultaneously, each shooting and killing the other. Then another of the railroad men, named Walter Crow, began firing at the remaining settlers with a shotgun, killing five men before fleeing. A mile and a half away, Crow himself was fatally shot in the back.

The event, known as the Mussel Slough tragedy, was seized upon by newspaper publishers and muckrakers as examples of the unbridled corporate greed of the monopolistic railroad companies. Popular opinion agreed. The surviving settlers from the shootout became folk heroes. After serving an eight-month sentence, they arrived home to cheering crowds of thousands. But regardless of who the heroes and villains were, the story highlights how people in the nineteenth century felt literally trapped by the railroad.

Railroads made California. Before transcontinental rail, the only way to San Francisco from the east was via months-long overland wagon routes—which offered the possibility of cholera, freezing, starvation, and attacks by Native American tribes—or taking a six-month trip by sea around the bottom of South America. Trains turned this epic journey into a comfortable week-long trip. As a result of their efficiency and mass accessibility, the railways opened the West to huge numbers of Americans, many of whom became Californians.

But trains created a whole new set of problems as well. They were hierarchically controlled. The train system was often accused of political cronyism, bribery, and ignoring public needs. Their design was linear. As a result, areas near train lines flourished, but much larger tracts of land not serviced by rail were left undeveloped and often inaccessible. Their scheduling was also inflexible, causing consternation among riders who chafed at arranging their daily lives around train timetables.

As a result of these shortcomings, railroads were rarely celebrated for their revolutionary contributions. In fact, they were more likely to be demonized as inconvenient, autocratic, and corrupt. People felt trapped and constrained by track-bound transportation.

Then the automobile was introduced. For many Americans, the ability to customize their mobility offered them incredible freedom. Instead of riding where the train took them, they could drive exactly where they needed or wanted to go. They could also set their own schedules. At the beginning of the twentieth century, the car was still a plaything of the rich. By 1930, there was an auto for every 1.3 households across the country.

The customizable technology quickly led to many uses you would expect. People traveled around much more of the country, taking road trips for fun. For the first time, large numbers of Americans went camping and explored distant mountains and beaches. Americans also lived farther from their jobs, in areas where they could afford more spacious single-family homes. As cars grew in popularity, highways came to dominate urban planning. In California, non-freeways became known as "surface roads."

The technology created a market for a variety of businesses, including gas stations, body shops, oil-change businesses, car dealerships, motels, and roadside diners. Retail businesses took advantage of more flexible delivery of products, workers, and customers to open up multiple branches. Manufacturing facilities relocated to cheaper, less crowded real estate with room for expansion. Farmers moved their wheat to centralized depots more cheaply, thus reducing the price of bread. Doctors began making house calls in the suburbs, as did salesmen. Regional managers visited far-flung operations more regularly.

The auto also allowed people an easy escape from the social mores and restrictions of households and neighborhoods. They were a mobile and private space away from chaperones, accelerating sexual liberation. The technology's impact was so widespread that a juvenile judge once decried the automobile as a "house of prostitution on wheels."

But the use cases of this customizable technology didn't stop with these more predictable adaptations. The automobile also created a network of new products, services, and jobs—some of which were unimaginable when Henry Ford's first cars rolled off his assembly line.

When autos opened up new tracts of land for development, they created a whole ecosystem of new industries. Housing contractors built single-family houses with yards. Landscaping businesses serviced them. Roofing companies fixed them. Real estate agents sold them. Then, shopping malls popped up to supply the residents. Drive-in theaters entertained them. Drive-through restaurants and pizza delivery services fed them.

Even less expected was the impact of autos on consumer goods. Large trucks and improved roads led to national distribution centers that, in turn, promoted innovations in packaging. For example, stackable flats of aluminum cans were developed to replace bottles as a more efficient use of space in the back of long-haul trucks. As a wider array of national products—from dish detergent to flour to mayonnaise—became available at large suburban supermarkets, they had to vie for consumer attention on crowded shelves. This competition prompted the creation of more eye-catching packaging as well as a new emphasis on brand management and modern advertising campaigns.

The point is that, while the immediate impact of the automobile's customization was massive, its repercussions went well beyond providing greater daily personal mobility to and from work. The auto put transportation technology in the hands of its users. Once the automobile was introduced, people found a whole new set of uses for it in ways that never happened with the train.

Today we are seeing the same thing happening in manufacturing with 3D printing. Like railroads, mass production was revolutionary. But it was also inflexible. Mass production achieves efficiency by producing huge volumes of the same thing. Out of necessity, these economics discourage customization. By contrast, with 3D printing unit costs are the same no matter how much you customize. At low volumes, 3D printing allows you to personalize what you make much more cost effectively than mass production. These economics, in turn, encourage customization and innovation.

CUSTOMIZED LIKE ME

Some of the early adaptations of 3D printing's customization have also been relatively predictable. Consider, for example, the Mini-Me. You stand in a photo booth to have a 3D image created. The file is sent off to a 3D printing factory. A few days later, you receive an exact six-inch replica of yourself. Another company has partnered with Mattel to print your face on a Barbie or Ken doll.

Design software company Autodesk has taken this type of personalization up a notch with an app called 123D Catch. Instead of snapping vacation pictures of yourself in, say, Florence, you can create a 3D file with your phone. When you get home, you can print out a statue of yourself in front of the Duomo. These are essentially high-tech tchotchkes—not disruptive use cases. But some people still find value in the design customization itself.

Nike is working on a more useful kind of personalization. Today, nike.com has a customize tab that allows users to personalize their own shoes—up to a point. NIKEiD allows you to choose from a range of colors and finishes or materials on every visible part of the shoe, from the uppers to the tongue to the sole and laces. You can also put initials or a short message on the back. Soon however, 3D printing will customize the shoe's actual performance.

Professional athletes already run in bio-personalized shoes. Engineers record data from the bodies of athletes moving on a controlled track. After analyzing their foot strikes, stride length, and other metrics, the designers can print out soles specially designed for that athlete's foot. Nike has also done research on how to build a cleat that allows athletes to accelerate immediately on grass out of a standing position. The result of this advanced engineering project has already been seen by millions. The 3D printed Vapor Carbon cleats were worn by both the Seattle Seahawks and the Denver Broncos in the 2014 Super Bowl. Now Nike is focusing on bringing this higher level of customization to everyday consumers.

Nike describes its Flyknit technology as knitting a sweater on your foot. The shoes are made in one unit; there is no assembly necessary except the sole. This makes it a perfect process for eventually 3D printing soles for individual feet. In late 2015, COO Eric Sprunk announced that the company will soon allow consumers to design their own shoes to the unique specifications of their bodies. Customers will print out their shoes at a local Nike store. "That's not that far away," Sprunk said.

Several companies are competing to create another kind of customized high-performance apparel. Military personnel on missions need clothing that allows them freedom of movement. But sometimes they also need clothes that can take a bullet for them. But most traditional products that provide this level of protection are also bulky and uncomfortable. 3D printed body armor that mimics fish scales may eliminate this problem. Fishes scales protect them from predators, but are also extremely flexible. The military has 3D printed a prototype of a suit that places overlapping stiff plates on a flexible underlayer. The armor will offer better protection than Kevlar while allowing the wearer to move relatively unencumbered. More importantly, the armor will also be printed to custom-fit the wearer's body and preferences. Depending on missions, users can design suits that trade flexibility for greater protection. The range of uses extends from a hidden sniper resting in cover to special forces in armed combat.

A competing 3D printed body armor concept is being designed and prototyped for the U.S. Army by Hollywood FX studio. The group is studying how to best combine various advanced materials in redesigned suits that have fewer seams for added comfort and strength. The printed product may even have embedded ballistic sections printed on a single piece of clothing.

GE also has a product that harnesses nature—but for a completely different use. The company 3D prints customized turbine blades for power-generating windmills. These blades function very differently depending on their environment. In low wind speed areas, for example, they face different challenges in performance and maintenance than in higher-wind areas—or areas with changing wind directions. Custom design in blade pitch or construction materials makes for a more efficient and durable turbine.

TOTAL VALUE

Fish-scale body armor or shoes that perfectly fit your feet are very cool. But, still, you may have predicted that 3D printing would be used for that kind of personalization. You also might wonder how 3D printing's customization is going to have a broad impact on traditional manufacturing via Mini-Me statues. After all, it still costs a lot more to print customized refrigerators and cars than to mass produce them. Why would companies or consumers choose 3D manufacturing?

In even posing that question we are revealing one of our conceptual stumbling blocks. In the age of mass production, we are used to thinking of "costs" on a per-unit basis. But this equation changes as completely new uses develop. For example, if there is an iPhone 25, it will likely be mass-produced. But what if 3D printing creates fitted, customized wearables that track a whole variety of personal metrics and serve our communication needs hands-free? The iPhone would suddenly seem like a relic—like dragging around a phone booth. The new personalized wearable can be cost-effectively manufactured only using 3D technology. Suddenly we have a use case where the old economics of marginal costs and production speed go right out the window. Mass production's advantages are nullified. To understand the disruptive capacity of 3D printing we have to stop talking about unit costs and start looking at total costs.

Traditional manufacturing may always be much faster and cheaper on the factory floor than 3D printing. But remember that the process sacrifices customization and complexity for scale and simplicity. Mass production of low per-unit cost items has hidden expenses: constricting design constraints, huge outlays of initial capital that discourage popular innovation, and an inability to customize.

As 3D printing encourages totally different customized use cases, products, and industries, mass production's shortcomings will become increasingly glaring. With each new use case, a new economics of manufacturing emerges. Eventually 3D printing's advantage in total costs will begin to overwhelm mass production's unit costs.

FORD EXPLORES THE SECOND WAVE

The Ford Motor Company is already developing new use cases for 3D printing as part of its industrial production. Seven years ago, the company experienced a major problem during the development of the 2011 Explorer. The new Explorer was rede-signed to be a showcase of Ford's research into haptics—how drivers' sense of touch affects their overall impression of products. Engineers had installed knobs, buttons, dials, and switches that felt solid, smooth, and crisp. The reassuring tactile sensation of the controls conveyed a feeling of overall quality to the customer. But an issue cropped up during the pre-build process. Test drivers had noticed a "Noise Vibration Harshness" during braking. For the vehicle to vibrate even slightly during braking would be a disaster. In these models, it was much worse. Stepping on the Explorer's brakes caused the vehicles to rattle and grind.

After troubleshooting the SUVs, the brake team decided there was a problem with the rotor design. Brake rotors, or discs, are the circular, pan-shaped pieces of metal that the brake pads press up against to stop a vehicle. The parts are also surprisingly complex. "If you were to look at the rotor on your car, it wouldn't look so complicated," said Harold Sears, a chief engineer with Ford's additive manufacturing team, "but there is quite a bit of complexity in the design." Even a slightly warped rotor will vibrate throughout the vehicle.

Traditionally this kind of error creates a serious problem for an automobile's deadline. The engineers would quickly design their best-guess fix on a replacement rotor. Then they would send off the design to an outside vendor and wait. The new rotor mold would take weeks or months to make. During this time, the engineers—and the whole Explorer production team—could do little else to fix the problem.

The rotor was critical for performance and safety. It absolutely needed to be correctly redesigned. Unfortunately the engineers' production schedule and budget didn't leave leeway for experimentation. The traditional design process encouraged conservative simplicity. The engineers were not going to take any risks with innovation. They only got one chance. They had to pick a design and stick with it.

In a sense, Ford's brake team faced the same dilemma as NASA's scientists during the Apollo 13 mission. An oxygen explosion onboard crippled the spacecraft halfway to the moon. The engineers on Earth plotted a new course that would use lunar gravitational pull to shoot the spacecraft around the moon and back home. This was a stunningly complex engineering problem. They had to deliver the astronauts across hundreds of thousands of miles of space to a planet moving at eighteen miles a second while accounting for limited electricity, decreased oxygen levels, and damaged exterior tiles on the spacecraft. And there was no Plan B. Of course, the automotive engineers' problem was much lower risk. What both groups had in common was that they only got one shot at solving a demanding engineering problem.

The dilemma faced by Ford's rotor team was far from isolated. In fact, similar issues have plagued industrial engineers for decades. Engineers are notorious for their perfectionism. There is an old joke in manufacturing that an engineer and an artist are the two people who never stop working until you pry the pencil out of their hands. But the realities of traditional mass production, tight schedules, and budget constraints don't permit an ideal work environment. There are two unfortunate results of these pressures. First, a high percentage of designs are, in fact, flawed and sent back for costly redesign. Even worse, some designs are flawed but too expensive to fix. The product goes to market with its noncritical design errors. Engineers simply have to grit their teeth and live with the result until it is cost-effective to do a redesign.

While these constraints are common in industrial engineering, not all designers have to put up with them. When trying to optimize a website, for example, web designers do not use their best guess and stick with it. Instead, best methodology dictates that coders create many different solutions. They can even integrate customer feedback into these designs. Do customers respond more to the blue picture or the yellow one? Does tag line A or B result in more click-throughs? Unlike industrial engineers, web designers solve problems through a continual iterative process. By comparing results from all this data, designers can mathematically converge on the absolute best answer.

This technique, frequently called split or A/B testing, is only possible because of the Internet's unique capacity for customization. If a print-only newspaper had a design problem, it would need to be fixed before the issue was printed, a very quick turnaround time. Like auto engineers, print designers don't have the luxury of trying dozens of different options, testing them live, and then converging on the best solution. They have to pick the single best approach that could be developed in the short turnaround time. But on a news website, thousands of customized variations of content and design need to be relatively quickly and cheaply generated and tested.

Until recently, this sort of Internet-enabled design customization had never been possible in the physical world. Ford's engineers were never able to work with multiple production-quality customized versions of the rotor. But, after examining the Explorer's brake rotor issues, this was exactly what the leader of the brake team asked for.

"We have to test multiple versions," he said bluntly. "And we need them quickly." The engineers wanted to produce eight different rotor designs so they could diagnose the problem and figure out the best solution—all while staying on their tight production schedule. Because they had access to metal 3D printers, that's exactly what they did.

The engineers printed multiple iterations of the rotors in-house within days. The parts were produced at less than a tenth of the cost of traditional prototyping. Inside of a week, the brake team had tested and successfully redesigned the part. The test driving could continue. 3D printers saved Ford months of lost production time and hundreds of thousands of dollars. In 2010, the redesigned Explorer was delivered on schedule. Soon thereafter it was named North American Truck of the Year.

Engineers at Ford don't just use 3D printing for design rescue operations like the brake rotor. They also improve their designs from the beginning. Take, for example, intake manifolds, the parts that direct air down into each cylinder. Intake manifolds look like a lost relative of the trumpet. But though they are a relatively simple collection of interconnected tubes, they can have huge impacts in performance. They also significantly vary in design from vehicle to vehicle.

The traditional design process for manifolds is clunky, expensive, and slow. "For one intake manifold," said Sears, "it takes from ten to eighteen weeks. Depending on complexity, you could be looking at anything from $50,000 up to $200,000 in cost."

3D printing testable intake manifolds allows engineers at Ford an enormous amount of freedom that previously never existed. Like web designers A/B testing designs online, they can print and test multiple iterations. This new use case is called Rapid Design Iteration. Not only is it a superior way to design a part, but Ford ends up saving time and money on the whole process.

"The 3D printed manifold takes about two days to fabricate and maybe another couple days to do the final prep and assembly," said Harold. "So we'll say within four to five days, that thing is on the engine and ready to run. At a cost of a couple thousand dollars." These figures are astonishing. Three months versus four days. $200,000 versus $2,000. To make the exact same piece.

Rapid Design Iteration has already changed manufacturing. And it is only possible because of 3D printing's capacity for low-volume customization. Mass production can't compete. And as Ford has become more familiar with 3D technology, they have discovered still more use cases, including some components for the final product. In fact, end-use parts are now the real drivers of growth in 3D printing—going straight at mass production's core use. In 2014, U.S. sales of the industrial-grade 3D printers used to make end-use parts were already one-third the volume of sales in the well-established industrial automation and robotics sectors. Some projections have that figure rising to 42 percent by 2020. For Ford and other companies, 3D printing's advantage in total costs is delivering real value and solving real business right now.

THE ISLAND OF MISFIT PARTS

In 1994, soon after he began working at Stratasys, engineer Jeff Hanson was working on a design file a client had sent in. The piece of geometry on his computer screen began to look strangely familiar—some sort of joystick controller.

He asked around the office and found out that the client was refurbishing a B-17 bomber, the massive "Flying Fortress" used by the U.S. Air Force in WWII. The body of the vintage plane was in good shape, but finding replacement parts for the interior was practically impossible. Most of the planes had been shot down or destroyed in combat half a century earlier. By the 1990s, only a few dozen B-17s remained anywhere in the world. No private market for the components existed. The original manufacturer, Boeing, couldn't help with a replacement. The joystick was the ultimate dead part. But the then-new technology of 3D printing offered a solution. The clients created a design file of the joystick and sent it off to Stratasys. Not only was Hanson able to quickly print a functional replica, but it was much cheaper than anything that could have been produced in low volumes through traditional mass production.

More recently, Hanson found himself in Jay Leno's garage, the massive warehouse owned by the comedian and car enthusiast. The garage's collection of autos is so extensive that, in 2015, it became the basis for a CNBC TV series. Keeping all these vintage, antique, and specialty cars running is a demanding task. When a part gives out, it can be nearly impossible to find a replacement in working order. But, using reverse engineering, Stratasys prints customized parts to get the cars back on the road.

Today, the same printing technology can also make household items. Imagine you are missing a part like a plastic shelf bracket on your refrigerator. The fridge still works fine, but the shelf can't hold a gallon of milk. Because it is ten years old, the manufacturer considers the fridge obsolete and won't service it. In the past, this situation resulted in an imperfect fix—or buying a new appliance. Today, you can have replacement parts printed out locally, in the exact number you need. Lowe's, for example, is piloting a 3D printing service. You bring a part in and they scan it and print as many replacements as you need—in plastic, metal, or ceramic.

Over two decades after Stratasys printed out the joystick for the B-17 bomber, 3D printing is still a very expensive way to make parts on a per-unit basis. But the technology's capacity for low-volume customization is creating new use cases, like appliance-saving obsolete parts. And that gets us back to unit costs vs. total costs.

Let's say a mass-produced fridge bracket costs 20¢ while a 3D printed version costs $20. The price is 100 times greater, but the economics are no longer about unit costs. The total cost includes the ultimate value of the part. If there are no longer any of the mass-produced brackets available, then the value of the customized bracket is the continued use of the refrigerator. The value of the customized joystick is a completed vintage WWII bomber. The value of a customized wheel rim is a fully restored classic car. Again, when you only need a few customized pieces, mass production can't compete on total costs.

THE NEED FOR SPEED

NASCAR teams frequently need to make adjustments or repairs to their vehicles between events. Because races are typically held every Sunday, any changes or repairs between events need to be made in a very short window. The teams have less than a week to design, test, manufacture, and install brand-new high-performance auto parts. With traditional mass-production technology, these time frames are unthinkable. It would take at least weeks to get a single part. This is why Joe Gibbs Racing uses 3D printing to accelerate the production cycle.

In one example, a car's front tires kept overheating and exploding on the track, causing the car to spin out of control. The engineers needed to get more air passing between the wheels and the engine to cool them off. On Sunday evening, they analyzed data from that day's race. The next day, the engineers designed a prototype part designed to fix the problem. They printed it out in about four hours and put it on a car. It didn't fit properly. No problem: The team redesigned it with corrections and printed out another iteration that solved the performance issue.

Up to this point, the engineers had just been using 3D printing as a rapid prototyping technology, much like Ford engineers with the intake manifolds. But once they had settled on a design, they took advantage of another rapidly expanding use of 3D printing. The end-use part needed to be manufactured from carbon fiber, an extremely lightweight, heat resistant, and strong material. Their 3D printer couldn't handle carbon fiber. But the printer could print out a temporary mold that could be used to make the part. As a result, the engineers didn't have to send their design off to a shop and then wait weeks for a heavy production mold to come back. They could 3D print the temporary mold in a few hours. Temporary 3D molds are typically used for making just a few dozen parts, not tens of thousands. But the team didn't need more than a few. After printing out the mold, the engineers poured in carbon fiber and waited for the material to set.

The part was installed on Wednesday. The car was loaded onto a truck later that day and shipped off to the next Sunday's race. This high-end, customized part was designed, tested, manufactured, and installed in just days using a computer and a 3D printer about the size of a large refrigerator.

Another example is a Georgia Pacific engineer who contacted our company, Fast Radius, a year ago. He urgently needed forty brackets to be used in a gear mechanism. He uploaded his file and hit the print button on our website at 3:30 on a Monday afternoon. At 9:15 the next morning, the parts were delivered to him.

Once again, the economics of 3D printing are often not about unit costs, but total costs. Joe Gibbs Racing relied on a new use case, 3D printed temporary molds, to get the car back on the track within a week. Georgia Pacific didn't have to interrupt a manufacturing line waiting for parts. Neither result would be possible under the constraints of mass production.

FACTORIES ON LAND AND SEA

The Danish cargo line Maersk's largest ships are the size of four football fields and are stacked with enough twenty-foot cargo containers for eighteen thousand tractor-trailers. The vessels spend months at sea, but their schedule is tightly monitored. So when something needs to be repaired, the ships can't afford to stop at the nearest port. Inevitably, though, something that big and heavy in the open ocean will experience maintenance problems. Traditionally all the options for getting replacement parts onboard, couriering them via plane or boat, have been expensive. The ships carry a limited amount of critical replacement parts, but space and weight are at a premium. Recently, Maersk developed a new solution to their problem: The company installed metal 3D printers onboard. Now routine maintenance issues can be fixed with no disruptions to operations, or added expenses.

Similarly, the construction giant Caterpillar sets up 3D printers at its work sites. From these mobile printing stations, it can print out all sorts of low-volume but critical parts like track links for a tractor. This capacity dramatically increases the uptime on their projects. In 2015, the company saved hundreds of thousands of dollars by using 3D printing versus conventional manufacturing. Caterpillar has become so adept and confident in 3D printing as an on-site repair service that it is now offering a new product: uptime insurance that guarantees their equipment will either work or be repaired quickly.

Most commercial quality printers are about the size of a large appliance, such as a dishwasher or refrigerator. They can fit in an office or on a work site. As a result of their mobility, 3D printers offer another kind of customization. They provide mobile, distributed manufacturing. Again, the per-unit cost of 3D printing is high, but that is not how we determine value. For Maersk, the value is staying on schedule. For Caterpillar, the value is increased uptime. When working in large construction sites in developing countries like Nigeria, you may even have to drive past armed rebels to get a replacement part in place, making the value of an on-demand printed part even higher.

3D's capacity for low-volume customization has already led to products you would expect. But it is also leading to brand-new use cases and business models. No one expected the automobiles' capacity for customizing transportation to impact the design of Tide boxes. Similarly, when 3D printers were just churning out plastic prototypes, no one imagined they would be used to make customized parts for NASCAR teams or windmill turbines. What's so exciting is that, because the technology is customizable, people will develop more and more uses for it as they get their hands on it. Caterpillar engineers say that once they had the 3D printer on-site, people quickly found new uses for it, including making specialized tools.

In each case, companies are choosing 3D printing's advantage in total costs versus mass production's lower per-unit costs. Mass production is fast and cheap, but only when you are producing massive amounts of things. By comparison, 3D printing has fewer and cheaper steps prior to production. It allows engineers to create multiple design iterations more rapidly and cheaply. A/B testing can give feedback on real-world performance before parts are produced in huge volumes. 3D printing of end-use parts is a cost-effective way to re-create obsolete parts. The technology is superior for making parts on tight deadlines. Critical repair pieces can be manufactured on unique job sites around the world. Finally, parts printed with customized design have a whole range of unique values, from personalization to performance characteristics.

The real-world benefits of customization have grown steadily over the past ten to fifteen years. To remain competitive, reduce total costs and increase profitability, a growing number of companies will be compelled to take advantage of these opportunities. But there is also an accelerator to this dynamic: Opportunities for customization will grow exponentially as 3D printing flips modern design theory upside down.

5

DESIGN WITHOUT LIMITATIONS

**Design is not just what it looks like and feels like.
Design is how it works.**

—STEVE JOBS

ON EASTER SUNDAY OF 1941, a chemist hunched over a table watching roaches die. A research employee at USDA headquarters, Lyle Goodhue had just sprayed the insects with mist from a bulky five-pound canister. One by one, they stopped moving. Within ten minutes, all thirty of them lay on their back. He let out a scream, danced around, and then ran out to his car and drove home to tell his wife. Goodhue had invented the first commercially viable aerosol spray can.

The next day, Goodhue had to present the invention to his bosses. They were considering cutting funding for the aerosol program. Again, the spray container performed well—killing dozens of roaches in a display for his research director. His boss was only mildly impressed, but decided not to eliminate the program. Goodhue and his colleagues took the canister back to their lab and continued working on it. A few months later, Goodhue's invention was awarded a patent. Aerosol technology was not new at the time. Goodhue's achievement was being the first to create an easy-to-use can with an accurate trigger mechanism. He imagined the invention being used as a convenient home insecticide or cleaning agent. But its biggest impact came far from the United States.

In 1942, the Japanese took the Philippines, handing U.S. forces their worst World War II defeat in the Pacific. The Americans retreated to Australia to assess their loss. One problem was supplies. Because of the war in Europe, they had been unable to get all the weapons they requested. Additionally, their tens of thousands of Philippine allies were poorly trained. Most had never fired a weapon before the Japanese invasion. But the U.S. Army's biggest single problem was mosquitoes. Thousands of troops were sickened or dying from malaria. Goodhue's design, known as a "health bomb" at the time, was the solution. Soldiers used over 40 million of these canisters in WWII, spraying the inside of tents, airplanes, and other enclosed areas. Goodhue's insecticide delivery device is credited with saving thousands of American soldiers.

Goodhue's design work was impressive by itself, but it is truly amazing when you realize the restrictions he was working under. Before he even picked up a pencil at a drafting table, he had to deal with three critical constraints. First, he could only use limited complexity. He needed to create the simplest container and dispersion trigger possible. If not, the mass-production methods used to manufacture them for soldiers would run up against cost or feasibility issues.

Second, he could use only existing materials. Certainly he would have liked stronger or more lightweight material to accomplish his objective better, but he had to work with steel and aluminum—there weren't many other options.

Third, he could use only his brain. His brain was impressive—he had a nearly perfect memory—but it was capable of imagining only so many functional possibilities. Typically, designers start with three to five possible designs in their mind. Then they begin working and testing those concepts. Innovative engineers like Goodhue hit upon a design that gets the job done. But there were thousands of possible alternative aerosol spray can designs. He just couldn't tap into them.

What's shocking is that, even with all our technological advances since 1941, most designers contend with the same three problems today. Imagine if we didn't have any of these constraints. What if we could build anything we wanted without worrying about complexity, limited materials, or humans' inability to generate every functional option?

PUTTING PILLS IN THE FAST LANE

In late 2015 in a pristine pharmaceutical lab, these constraints literally melted away. For the first time in history, the FDA approved a 3D printed pill called Spiritam. The medication was designed and manufactured without any of the constraints that have dogged designers and engineers since well before Goodhue's time. The future of design had arrived.

Aside from Spiritam, all medicines and pills are mass-produced in huge factories at high speeds, just like shirts and cars. They use the simplest shapes, limited materials, and rarely innovate on design. This presents multiple problems. First, drugs only come in standardized quantities of milligrams. This means heavily prescribed patients take many pills a day. They often have multiple dosage times, which they forget. Other patients have to split pills at home with a knife—an imprecise measurement. Finally, doctors are limited in what they can prescribe. Their scripts have to be measured in multiples of 250 milligrams, regardless of a patient's body chemistry. Largely as a result, many people simply do not take their medicine. Enormous sums of money and time spent developing life-changing drugs are effectively squandered by design limitations on the delivery device. This is bad for the drug companies—and much worse for the patients.

Spiritam is not a new kind of medicine. The active ingredient, levetiracetam, has been used to treat epileptic seizures for years. The pharmaceutical company Aprecia picked it, in part, because the patent had expired. But the pill packages the old medicine into a new kind of delivery system. By 3D printing layers of powder into a water-soluble complex matrix, the drug instantaneously dissolves in the mouth. The active ingredients are immediately absorbed into the blood stream. This extra speed helps suppress sudden onset seizures. The pill is also much more comfortable to take. Previously, the minimum dosage for levetiracetam was a bulky 500-milligram tablet. There is no need to reach for a glass of water to swallow a 3D printed pill. Aprecia has even developed a process that 3D prints pills as quickly as traditional factories can mass-produce them. Next, the company will be printing personalized drug dosages. As a result, patients who take multiple pills daily could take just one. Simply by eliminating long-standing design limitations, 3D printed pills can make existing medicines more effective.

The market for pharmaceuticals is incredibly profitable and rapidly expanding. So why has it taken companies so long to create a delivery device with all the advantages of Spiritam? Partly because 3D printing has only recently gotten to the point where it can print with such complexity. Second, the material—a printable drug powder—is relatively new. The third reason is design-related. No human engineers could imagine the complexity needed to create Aprecia's lattice structure. Instead, the matrix design is uniquely generated each time by a computer based on the specific prescription and dosage.

Aprecia seems simple. It's just a different way of making a pill. But look closer. The pill breaks the three constraints designers have struggled with since well before Lyle Goodhue's aerosol can. It leverages unlimited design options to create completely customized materials that are produced in completely customized geometries—no matter what your dosage, you get a single pill that melts in your mouth. Aprecia is part of the leading edge of a new 3D printing–enabled design ethos. Ultimately, 3D printing is not about making the same things in a cool new way. It is about enabling an entirely new approach to design that allows us to create things beyond our dreams. Let's look closer at how boundless complexity, novel materials, and computers that generate thousands of possible designs are already beginning to change design in medicine, aerospace, electronics, and even glass making.

COMPLEXITY FOR FREE

In 2015, the first jet engine with a 3D printed part took flight. Designed by GE Aviation, the part solved a chronic maintenance problem. During flight, ice built up on a sensor mounted on the outside of the engine. Eventually, chunks would break off and slide into the engine's compressor, damaging it. GE engineers fixed the problem with a series of precise and delicate geometries that stopped ice buildup before it happened.

The result was a curious component. The part looked like a cobalt-steel flower with a squished stem and two protruding stamens. In other words, it looked like a nightmare to mass-produce. Traditional manufacturing always prefers the functional part with the simplest geometries possible. It's faster, easier, and less likely to result in manufacturing errors. 3D printers don't care about complexity. The additive process, building layer by tiny layer, produces extremely complicated objects just as easily as simple ones. There is no cost for adding exotic geometries to your design. As a result, 3D printing encourages a new kind of design: using as much complexity as necessary to optimize your part.

In the case of the sensor housing, the designers took full advantage. Liberated from design constraints, GE's engineers had created a part so complex that it simply could not be traditionally manufactured. To 3D print the parts was expensive on a per-unit basis versus a mass-produced alternative. But the complex design had enormous efficiencies. So GE printed four hundred of the parts and began retrofitting them on their engines.

The decision to print this small bit of housing represents a huge shift both in design as well as how companies use 3D printing. In the 1990s, 3D technology was almost exclusively used for rapid prototyping parts that would later be mass-produced. But, as engineers were given more design freedom by the technology, they began designing parts that couldn't be manufactured traditionally. A new ethos developed. Why not use 3D printing's cost-free complexity to create new kinds of functionalities by building parts that were never before possible?

HOLISTIC MANUFACTURING

As much as any other company, GE has embraced this new approach to design. Its innovative fuel nozzle system we mentioned in chapter 1 is extremely complex. The part can only be made with 3D printing. The company has committed to manufacturing forty-five thousand of the parts every year. This is hardly low-volume production, but GE still expects to save 75 percent on the cost of manufacturing versus mass production. How is this possible?

Once you can design without restrictions on complexity, you discover all kinds of new advantages. Some of them are functional: Spiritam delivers dosage rapidly; the housing sensor prevents ice buildup. The design of the nozzle system also enhances efficiency and functionality. But it does so by reducing the number of components. The previous fuel nozzle system was machined out of twenty-one separately cast pieces. Each part was produced by a part supplier, shipped to a central assembly plant, and then put together by hand. The new system is printed as a single part. Holistic manufacturing, as it is known, uses design complexity to create what is ultimately a simpler part.

One advantage of holistic manufacturing is the lack of welds. In traditionally built fuel-injection nozzles, the welds that hold the multiple parts together create air friction that compromises both the internal air flow and, as a result, the internal temperatures as well. With 3D technology, engineers can print tiny wind tunnels that flow directly along the path of the fuel. When your fuel is being blown out at 1,500°F ten times hotter than the melting point of the parts it flows through—it's hugely beneficial to keep both the fuel as hot as possible and the temperature of the parts as low as can be.

Second, because the part is printed as one piece, GE didn't have to ship these parts from various suppliers around the country—or world. GE's Christine Furstoss says that products with "fewer frequent flyer miles" represent a significant savings for the company.

Mass production may be able to produce tens of thousands of individual parts more cheaply than 3D printing. But once the total cost of manufacturing—including shipping and assembly—is added, 3D printing gains a cost advantage. "The real power of additive is taking six parts and designing them into one," says Greg Morris, the additive technologies leader at GE Aviation. "You can create geometries that you just can't make any other way."

The enormous appeal of more complex design that limits assembly work has driven investment in ever-larger 3D printing equipment. The Department of Defense, Lockheed Martin, Cincinnati Tool Steel, and Oak Ridge National Laboratory are partnering to develop the capacity to print most of the endo- and exoskeletons of jet fighters in one piece. Instead of printing just one superefficient part of the engine, "big area" 3D printing machines will manufacture the body, wings, internal structural panels, embedded wiring, and antennas. Instead of working inside an appliance-sized box, these massive machines use a computerized gantry with controls that move the printer over a space of dozens of feet.

The basic capacity is already in use. Aurora Flight Sciences holistically prints the exoskeleton of the drones that fly in the skies above Iraq and Afghanistan. The company manufactures the outside of entire aircraft, sometimes with wingspans greater than one hundred feet.

PRINTING THE IMPOSSIBLE

HRL Laboratories calls its "microlattice" the world's lightest metal. The material actually looks more like a stainless steel scouring pad than any common metal. Built out of a series of interconnected hollow tubes, the material is almost entirely porous and incredibly flexible. When compressed or bent, it flexes instead of breaking. It also absorbs force remarkably well. HRL engineers claim that an egg wrapped in a small microlattice box could be safely thrown out of twenty-five-story window. This is the sort of super material you end up with when you start designing with none of the traditional restrictions. The engineers' design goal was to remove as much material as possible. They succeeded: The microlattice is 99.99 percent air.

Boeing is looking at developing the microlattice for use as walls or doors in aircraft. Not only would the material be incredibly hard to break, it would provide enormous reductions in weight, fuel cost, and environmental impact. The material's amazing functionality is purely a result of its very complex design. Whatever commercial applications develop, they will have to be 3D printed. It's impossible to make the microlattice any other way.

3D printed windpipes are another example of designing without fear of complexity. Windpipes, or tracheas, are very complicated and critical body parts. They have the vital job of carrying our breath from the larynx down to the lungs. Tracheas need to be strong enough to withstand the extreme pressure changes of coughs and sneezes. But they can't be too rigid, either. They must bend in order for the neck to move freely. There is also no room for error with a trachea; the bio-customization has to be perfect. If it doesn't fit, the patient can't breathe. The trachea is a body part with very complex, demanding specifications. Until recently, making replacements has been impossible.

But over the past several years, scientists have used 3D printing to create the first functional trachea. Using a design that makes use of complexity and novel materials, they have planted stem cells onto a 3D printed support structure. The living object grows into a customized bio-replacement. To date, surgeons have successfully transplanted tracheas into multiple patients. These recipients include a two-year-old Canadian girl born without a windpipe, a birth defect that is otherwise fatal. Her body fully accepted the printed replacement.

SOLVING NATURE'S RIDDLE

Another new material, graphene, has a breathtaking number of potential applications. It could be used to create unbreakable touch screens or upload a terabyte in a second. It could make batteries obsolete, filter salt out of ocean water, clean up radioactive waste, make tires that never go flat, and repair spinal injuries. But the story of the super material's discovery is just as remarkable as its potential.

Andre Geim is an eccentric Russian physicist with a frumpy hair cut who could be easily mistaken for Bob Denver of *Gilligan's Island* fame. In 1997, Geim used a magnetic field to levitate a frog. He later coauthored a paper on the earth's rotation with his hamster, credited as H.A.M.S. Ter Tisha. During 2003, he and his colleague Konstantin Novoselov hosted "Friday Night Experiments" at the University of Manchester. Geim saw the sessions as another unconventional approach to keep science exciting for his physics students. They encouraged students and professors to do experiments outside of their regular fields. One Friday evening, Geim decided to see how thin he could file a block of graphite. Geim and Novoselov ended up winning the Nobel Prize in Physics.

Graphite is a soft, black mineral that looks like a block of coal. It is made of thin layers of material that are just one carbon atom thick. The layers have relatively weak bonds between them and tend to separate and flake off material. Today graphite is mixed with clay and used in most pencils. It is also used in batteries and many other industrial devices and machinery. But Geim wasn't interested in any of those common applications. He was after a theoretically possible material so thin that it was essentially two-dimensional.

Geim asked a student to file down a block of graphite to as thin as he could. Using grinding tools, the student got it down to about a tenth the thickness of a human hair but couldn't go any farther. Chasing a hunch, Geim then put one of the ultrathin graphite flakes on a piece of scotch tape. He folded the tape over so that the flake was sandwiched in the middle. When Geim pulled the tape apart again, some weakly bonded layers of graphite peeled off. It was impossible to grind the incredibly thin material any farther, but Geim could remove layers bit by bit with the tape. He took one of the thinned-out pieces of graphite off the tape and put it on a new piece. Geim folded that over and pulled it apart. Each time, more layers of graphite peeled off. Eventually, Geim was able to isolate a single sheet of graphite just one atom thick. He had discovered graphene.

Graphene is a million times thinner than a sheet of paper, more electrically conductive than copper, two hundred times stronger than steel, and also incredibly flexible. It is also the lightest man-made material on earth. You could hold a sheet of graphene the size of an entire football field over your head with one hand. It is also transparent and almost completely impermeable to liquids. Geim and Novoselov's paper describing their discovery has become one of the most cited articles in science, and in 2010, they were awarded the Nobel Prize in physics for their work.

Once it was isolated, graphene was seen as a super material with limitless possible applications, but there was one very big problem. It was nearly impossible to manufacture, and far too expensive to make anything with. The possibility of commercial applications would be out of reach—if it were not for 3D printing.

In 2009, an entrepreneurial husband and wife team, Daniel Stolayrov and Elena Polyakova, developed a more cost-effective process for working with graphene. They began selling their product to universities under the name Graphene Labs. Soon they were supplying just about every research facility in the world. But they also realized that expanding the market for graphene would be nearly impossible without developing commercial products and applications. They needed low-volume manufacturing that could work with extreme design complexity. 3D printing was the only answer.

One of Stolayrov and Polyakova's advances so far is using graphene to create new kinds of composites. They have used graphene composites to make batteries, but eventually they hope to print electronic devices as a fully integrated unit.

The problem with printing such devices now is that they are made up of different materials with different functional properties. "Let's take a cell phone for a moment," says Stolayrov. "You'll see that it's covered with conductive glass. It has a strong body. Inside it has electronics that consist of some silicon materials and wires and printed circuit boards."

This may be obvious, but it is also a major hurdle to printing entire tablets or clocks or computers. To additively manufacture an entire product means coming up with materials that can perform each of the functions while also being printable. More importantly, they must be 3D printable together.

"Here we come to a problem," says Stolayrov. "Let's say that you want to have conductive and insulating materials within the same device. It will be produced by the same printer. But metal, used as a conductor, has a molten temperature far beyond that of plastics. If they were printed together, the metal would melt and destroy all the plastic components. So we need a replacement for metal with the same critical properties that can be 3D printed."

Stolayrov's solution is embedding plastic with graphene nanocomposite. Graphene is highly conductive. When the nanocomposite is added to plastic, plastic gains conductivity. This is more or less like adding red pigment to a plastic to give it redness. The new graphene "ink" has the conductivity of metal, so it can replace the wire, but it has the same melting point as plastic. Now conductive wiring and plastic can be printed together on the same printer to make the same device. In effect, you can embed functional wiring deep within an object in a single print.

It's unlikely that these materials would ever be used to print conventional cell phones, for example. The whole point of 3D printing novel materials is to design things with new functionalities. For example, the nanocomposites could be used in designing an ultra-lightweight, transparent, flexible, and unbreakable tablet.

At the Glass Robotics Lab at Virginia Tech, a team is developing a new materials science by studying the behavior of glass. 3D printers work with molten glass right out of a furnace. The glass is extruded through a nozzle on a large-format industrial robotic arm. The process looks sort of like printing with honey. As the glass winds down to the object below, it solidifies, becoming part of the design. But it also shows another behavior, what is called "transient coiling." This refers to the way that hot glass fibers rotate as they come into contact with other materials. By studying the self-organization of glass materials, the team can understand possibilities for new patterns and designs. They can use these natural capacities in future designs.

Glass is not a new material. But understanding how molten glass behaves can have the same net impact: creating previously unimaginable designs and products. The team is not simply creating a new super material like a microlattice or graphene to design with. It's developing a whole new process for working with materials. Instead of choosing a material for a design, they are collaborating with the material to optimize design. Eventually, 3D printed glass will almost certainly be used for advanced glasses lenses and fiber optics. As we better understand the properties of glass, it's impossible to say what kind of unbounded designs could emerge. We can, however, be pretty sure they will have functionalities we can't yet imagine.

TMI

The best feedback Lyle Goodhue got on his aerosol can came from dead roaches. But he was not alone in receiving limited useful information about his design. Other designers might hear from focus groups or reports of real-world performance: Did the seat wear out on a bike? It was hard to make well-informed judgments about what could be improved.

Today we have the opposite problem. There is an overwhelming profusion of information, not just from people, but from the products themselves. More and more objects have some sort of embedded sensor or electronics. For example, RFID chips let alarm clocks and thermoses tell warehouse employees where they are. Giant turbines describe their performance to software analysts. And new cars won't shut up, reporting on the mass of air entering the engine, the proximity to curbs and other cars, the position and speed of all metal moving components, and the alcohol content of the driver's breath. In short, we now have a mind-boggling array of real-time information from products. How can we design for this world?

The answer is a process called generative design, which is being pioneered by the software company Autodesk. Instead of picking a few ideas and working from there, designers will create the "DNA" for an object. They do this by telling a computer their objectives, what they are trying to optimize. Then the computer generates a near countless variety of designs, or "children." Generative design explores the whole solution space. It comes up with ideas that no team of designers could possibly imagine. Goodhue created one functional aerosol can design. Generative design could create thousands. Then, through a collaborative effort between human design teams and computers, the options are winnowed down until they decide upon the optimal design. "Increases in efficiency, performance, and quality will soon leap not by single digit percentages, but by orders of magnitude," says Mike Geyer, Autodesk's director of evangelism and emerging technology. "In the next five years we will see what was seemingly science fiction–type research quickly make its way into mainstream processes in almost every industry."

TOO MUCH, NEVER ENOUGH

The partition panel that divides cabins on the Airbus A320 is strong and lightweight. It's built out of a honeycomb composite.

It's been used for decades on Airbus' most popular plane. It's a very good design. So why is Airbus scrapping it? Because they can do something better now, using design without limitations.

To make this new piece, Airbus figured out exactly what their performance parameters were. Then they approached 3D software company Autodesk, which plugged the algorithms into a computer. Using generative design, the computer returned not five or ten possible designs, but tens of thousands of options, all of which met Airbus' goals. All that was left was to pick the optimal design. Airbus has also developed a proprietary powdered alloy called Scalmalloy to build high-strength, low-weight components like the partition.

The result is what Autodesk's CTO Jeff Kowalski calls a "bionic partition." 3D printed entirely out of Scalmalloy, it will save fifty-five pounds of weight while retaining all the required strength characteristics. Annually, the lighter partitions will save half a million metric tons of carbon dioxide, the equivalent of removing ninety-six thousand passenger cars from the road. The component has passed all of its initial tests and is now testing for flight readiness.

Until the 1980s people were still designing cars and planes on drafting tables. By the end of the twentieth century, computer-aided design was fairly prevalent, but designers were still coming up with all the ideas. The computers were passive participants. We had great inventions over that time, but all designers had to work under three constraints: simple geometries, limited material, and design ideas that entirely originated in their own minds. Now we are seeing these constraints begin to fade away.

In less than two decades, we have developed computers that can generate nearly endless iterations of designs. A palette of new materials is just beginning to rapidly expand. But realizing these designs would be impossible without 3D printing. The technology is what brings endlessly complex generatively designed objects into the physical world. In fact, the two are perfect compliments: Generative design is also the perfect tool to take advantage of 3D printing's capacity for complexity.

The end result of all these exciting possibilities is, to be honest, a bit frightening. Imagine having a nearly unlimited amount of geometries, a nearly unlimited amount of material properties, and a nearly unlimited amount of data. I was discussing this topic at a conference in 2015 when a CEO in the audience yelled out, "Wow, I guess I need much smarter engineers!"

He's absolutely right that the prospect is overwhelming from a traditional design and engineering standpoint. But there's a bigger issue. The amount of data and design possibilities will soon be overwhelming for *everyone*. It doesn't matter who you hire. Even an all-star engineering team won't be able to keep up with the millions of different data points that can be used to optimize a specific design. In five years, these equations will be so complex that no human will be able to optimize them. This represents another sea change in design theory. The best designers in the future will no longer be the ones who can conceptualize the best designs. They will be those who know how to ask the computers the best questions.

6

THE ELASTIC MANUFACTURING CLOUD

The secret of change is to focus all of your energy, not on fighting the old, but on building the new.

—SOCRATES

WHEN MALCOLM MCLEAN GRADUATED HIGH school in 1935, his family didn't have enough money to pay for college. It was smack in the middle of the Great Depression in Winston-Salem, North Carolina. But they did come up with enough money to buy a used truck. McLean drove the truck himself, drumming up business hauling goods around the South. Soon he was able to buy a dozen rigs and start a company with his brother and sister.

In 1937, McLean drove a load of cotton up to Hoboken, New

Jersey, for shipment to Istanbul. Following the six-hundred-mile drive, he sat waiting for days as the longshoremen loaded the cargo. McLean grew increasingly irritated. In those days longshoremen loaded items individually or strapped onto 4×4 pallets. The pieces would be painstakingly stowed in holds or lashed onto the deck of ships. It took so long that cargo ships typically spent as much time in port being loaded and unloaded as they did at sea. Because of the uncertainties of schedules, cargo arrived weeks ahead of sailing dates. Once there, it just sat on the dock or in the warehouse, leading to additional loss, damage, and theft. "There has to be a better way," muttered McLean to himself.

Soon thereafter, McLean sold his trucking business to buy a shipping company. He began experimenting with ways to speed up the transport of cargo. In one case, he drove entire trucks onboard his ships, but they ended up consuming too much unused stowage. Back in New Jersey nineteen years later, McLean realized his vision. He had converted a large cargo ship by building a railing around the deck. When it sailed away from Hoboken, fifty-eight truck trailers were stacked onto the deck. McLean had turned tractor-trailers into the world's largest shipping containers. When the ship arrived in Houston, the containers were hoisted off the ship with cranes and placed directly onto the beds of waiting trucks. The longshoremen didn't touch a single piece of cargo. The shipping container was born. McLean had created a better way to transport goods.

A decade later, McLean's company won a military contract to supply a large transport operation to Southeast Asia: the Vietnam War. Over the next six years, his company accounted for 10 percent of all U.S. cargo shipped to Vietnam. McLean could put more cargo on his ships more quickly than anybody else. His competitors were forced to imitate. By 1973, 80 percent of total cargo to

Vietnam was containerized. The first modern supply chain between Asia and North America had been established.

As the rest of the industry adopted containerization, Trailer Equivalent Units or TEUs became a global measurement of cargo. Container ships got bigger and bigger until the largest couldn't even fit through the Panama Canal. A huge growth in the volume of shipping followed. In 1983, the United States exported about 700 metric tons via ocean. In 1993, that number had grown to 885 and by 2003 it was about 1,170 metric tons. U.S. export by ship had increased by 70 percent in just two decades. And it all began with a man impatiently waiting at a dock for days, knowing there must be a better way.

TRAPPED BY LOWER COSTS

McLean's great success was eliminating friction. Just as Henry Ford's assembly line eliminated friction and unlocked low-cost production, McClean's containers unlocked low-cost global shipping. Like water breaking through a levee, global production quickly raced to a foregone conclusion, shifting production halfway around the world to the largest factories and the lowest labor costs. The result is today's amazing global supply chain that crisscrosses the world with raw materials and finished goods. Without McLean's container ships, low-cost, mass-production goods would be more expensive and fewer. But containerization also had a downside. Its ultimate state is long, linear global supply chains mired with inefficiencies.

Ironically, a system designed to reduce the cost of transport and chase lower wages has ended up creating new costs. Of course, shipping large quantities of cheap mass-produced goods from China to large markets in Europe or North America can result in lower per-unit costs. But it also leads to increased exposure to costly vulnerabilities. We may say we live in a globalized economy, but that doesn't rid the world of local problems. Natural disasters, political instability, trade restrictions, border delays, and currency fluctuations are among the dangers of shipping products halfway around the world. In 2014, over 80 percent of companies experienced at least one supply chain disruption. In 2015, political tensions and war in Ukraine and the Middle East caused significant problems. Look at the instability in the world today, and it's hard to foresee any fewer disruptions in the years just ahead.

Long shipping routes from Asian manufacturing centers also complicate manufacturing by adding weeks or even months of lead time between factories and end-point consumers or businesses. Shipping is a significant expense. But it also leaves no room for flexibility in delivery times. Short of air freighting goods, a container ship can't speed up to arrive days earlier. In fact, many ships now use slow steaming—reducing cost and carbon emissions, but adding days to delivery time.

These long, inflexible delivery schedules mean that businesses have to make manufacturing orders far in advance. For retail operations in particular, it's all but impossible to predict demand with pinpoint accuracy—and the process gets less accurate the further out they have to guess. When approximating, most companies overorder rather than risk upsetting customers by not having out-of-stock items. Overordering has two primary results: excessive quantities of goods or inventory that expires.

Our global supply chain has been made to serve our industrial mass-production manufacturing needs. It mirrors its successes, like lowering per-unit costs. It also has the same faults, like being slow and inflexible. And even with all the efficiencies that thousands of brilliant minds have squeezed out of it, the supply chain still creates an enormous amount of waste. One example: the 440 million metric tons of food that is unused every year. By weight, this annual waste is the equivalent of dumping out all the cow's milk bottled in the United States for the past five years. This nourishment never makes it to places where people are starving. And this massive loss is almost entirely—87 percent—due to supply chain problems. This is what an inefficient system looks like.

Finally, global supply chains are based on a capital economy. There are huge start-up costs to play in this game. This leads to the mass-production paradox: You can't enter a market if you can't produce in mass, and you can't produce in mass if you're just entering a market. This is an enormous barrier to new companies, one that locks out 99 percent of the possible innovators. But 3D printing is poised to offer solutions to many of these problems.

ELASTIC MANUFACTURING CLOUD

In 2003, engineers at Amazon began work on what they called "an infrastructure service for the world." Legend has it that the idea was an unexpected result of CEO Jeff Bezos' obsession with customer service. Amazon.com experiences massive seasonal spikes in retail activity. Bezos was determined that the website would never choke on peak capacity. This was a tall order. In the weeks leading up to Christmas, Amazon's servers were processing thousands of orders a minute. It would be technically possible to create enough capacity to absorb these maxed-out days. But the result would be a massive data and storage infrastructure that was underused most of the time. The extra capacity was like a cavernous airport that sat empty 350 days a year.

Whether or not the origin story holds up, by 2003 Amazon had definitely developed unparalleled storage and processing capacities. It held an inventory of over a million products. Selling everything from toilets to diamonds had given Amazon expertise at managing huge amounts of web-based data. Why not leverage this capacity and market it as a product in itself?

Three years later, Amazon Web Services launched. Since then, dozens of huge companies, including IBM and Netflix, have outsourced all or much of their data storage and processing needs to Amazon's cloud. The Pentagon and CIA store information in a high-security government-only cloud. When NASA's *Curiosity* robotic rover landed on Mars, it fed high-resolution images to the world in real time using Amazon's cloud. Amazon's Web Services has been more successful than almost anyone could have imagined ten years ago. In fact, the Web Services business is currently the most profitable division of the online retail giant.

Amazon's Web Services is used by companies of all stripes, but they have had a particularly historic and transformational value for entrepreneurs. Cloud-based infrastructures allow new companies to sidestep investing in expensive new hardware and complicated software installations right off the bat. Instead of worrying about buying $50,000 Cisco routers and servers before they can launch, they can pay for only as much data and processing as they use.

For some smaller players, these previous infrastructure investments represented a massive barrier to entry into the marketplace. These costs effectively left millions of would-be entrepreneurs out of the innovation value chain. Today, start-ups that outsource these data-infrastructure functions to Amazon's computing cloud no longer face a capital investment barrier. As a result, entrepreneurs and innovators can focus on developing their product. Amazon Web Services was a simple customer-service solution that had far-reaching implications. Maybe Jeff Bezos wanted nothing more than to make sure he never missed a sale. But his creation threw the doors wide open for everyone to participate, to innovate, to start a company.

Today's manufacturing supply chain has a similar set of prohibitive cost constraints. In fact, machinery, distribution, inventory, and warehousing logistics are often more complicated and costly than data-infrastructure expenses. Like any barrier, they also discourage entry for would-be manufacturing entrepreneurs. Until now, that is.

We already know that 3D printing excels in creating customized, low-volume goods. We know it can print locally. As a result it allows entrepreneurs to print on-demand without having to lay out huge capital expenses up front. They can also avoid shipping huge amounts of goods from China and then warehousing them until they are needed. And the result of all these capacities is one of the most disruptive impacts of the technology. 3D printing enables the manufacturing equivalent of cloud-based computing services.

Amazon's Web Services enables start-ups by streamlining and reducing the costs of data storage. 3D printing services reduce— or even eliminate—excess inventory costs. Cloud-based computing is a virtual environment with a pay-as-you-use model. 3D printing offers an unparalleled on-demand manufacturing capacity: tapping into a global network of production, as needed. Amazon got rid of the computing needs of a back office; 3D printing gets rid of the manufacturing needs of a warehouse.

Taken together, the Internet and 3D printing can offer huge advantages to the same entrepreneur. Need to raise initial capital? Internet-based Kickstarter has been a game-changing way for start-ups to secure seed money. With Amazon or other cloud-based on-demand networks, entrepreneurs can also save on initial costs by outsourcing their data management. This lets them use more seed money developing the end-use product itself.

After designing and funding an innovative product and selling it to their first eager set of customers, these entrepreneurs have to manufacture it. But before 3D printing, this next step was a major stumbling block. The entrepreneurs might need, for example, five hundred plastic cases to house their product in. But they have no manufacturing experience, own no machinery—and to their shock and surprise, they learn that it often costs $25,000 or more to make a single mold. Many start-ups, particularly those who begin life through a crowd-sourcing service like Kickstarter, just want to enter the marketplace with a few hundred dollars. They don't want to be stuck transporting or storing thousands of potentially excess or outdated versions. Mass-production economics don't help these entrepreneurs; the numbers only make sense if you are making tens of thousands of units. This is why traditional mass production and entrepreneurs are often a bad fit. In contrast, 3D printing's low-volume customization gives them exactly what they need: a relatively inexpensive way to launch their concept.

A few years ago, 3D printing company Stratasys helped a medical start-up do just that. Three doctors had developed a handheld defibrillator. They came to Stratasys to 3D print prototypes of their product. But during the process, they saw the benefits in using the technology for market testing and the end-use parts as well. The doctors did two hundred clinical trials and then made quick design changes. Once the design was complete, they successfully launched a small 3D printed run—2,500—of the defibrillators. 3D printing allowed the entrepreneurs to get to market fast and cheaply, rapidly iterating their designs. They avoided the infrastructure costs of traditional manufacturing.

Just as importantly, the fledgling business was able to mitigate the risk of ending up with tens of thousands of products in overstock. The nature of start-ups is to experience exponential growth—or go bust. Entrepreneurs with a manufactured product may have no idea whether they are going to need to produce 300 or 3,000 parts in the next few months. With 3D printing, you don't have to commit to decisions with huge lead times. Your parts are manufactured on-demand. You avoid excess inventory, machinery, and warehousing. It's an enormous benefit to new companies, but this is only scratching the surface of 3D printing's eventual disruption of the global supply chain.

On-demand production will almost certainly have an immense impact on enterprise-level manufacturing as well. A typical large auto parts supplier or aerospace manufacturer will often have one million unique parts or more. If 3D printing production can allow the transition of even one percent of these slow moving and low-volume parts (10,000), the business impacts will be tremendous.

But even with the new tools provided by 3D technology, such a huge transformation will not happen overnight. "Identification of these parts in real time is complicated, and requires analyzing production, shipping and transportation costs, tax implications, storage savings, and even environmental impacts such as reduced carbon emissions," according to Gil Perez, a senior vice president at SAP. To leverage the new capacities of on-demand production, SAP's back-end systems will need to be fully integrated into on-demand service providers—like UPS and Fast Radius.

"The adoption to an on-demand production economy will be gradual and will take time, perhaps five to ten years," said Perez. "At SAP we are committed to accelerating the digital transformation of businesses globally by providing our customers with the ability to leverage the latest on-demand manufacturing technologies in their business."

DIRECT DIGITAL MANUFACTURING

Philips Healthcare is the medical supply subsidiary of the massive German manufacturing conglomerate Philips. In the mid-2000s, demand for its products began to expand rapidly. The company was receiving 10,000 orders a day and held over 200,000 items in stock. During this rapid growth, its supply chain became increasingly complex, expensive, and, ultimately, overwhelmed. It couldn't make 10 percent of its deliveries due to warehouse capacity constraints. 40 percent of its returned parts were reparable, but getting them repaired and back in the field cost-effectively was another serious management challenge.

In 2006, Philips Healthcare commissioned UPS to implement what is called an advanced supply chain. The shipping and logistics company stocked warehouses full of replacement parts in its three stocking hubs. It stored smaller inventories in fifty dispersed "spokes," as well as using one thousand global pickup and drop off points. Today, UPS is able to get spare parts to Philips Healthcare anywhere in the world in twenty-four hours or less, guaranteed. The hub and spoke model also reduced inventory by 30 percent while increasing delivery reliability by 95 percent.

UPS' hub and spoke approach solved Philips' problems in two ways. First, it created an interlinked global network instead of a mishmash of local providers. But, second, its operations were still locally dispersed through its spokes. The results were impressive, but there was a trade-off. For the system to work, Philips had to invest in manufacturing tens of thousands of spare parts and then warehousing them around the world. Inevitably, some of these parts will be urgently needed in other parts of the world—or go out of date before being used.

Imagine a 3D enabled alternative. What would happen if UPS installed a hub and spoke 3D printing capability right on top of its existing hub and spoke delivery network? The hubs could offer a full range of services for repair and light manufacturing. The spokes could have very specific equipment, most likely certain types of printers and materials. But neither the hub nor spoke would have to hold huge amounts of inventory for anything that could be printed locally.

UPS' existing advanced system created significant efficiencies. But on-demand manufacturing takes this to the next level: dramatically reducing excess inventory. By replacing dispersed warehouses with local sets of industrial 3D printers, you could get the same quick turnaround capacities UPS offered—but without any overstock. Nor would UPS have to urgently, and expensively, ship 3D printable replacements to meet promised deadlines. As costs for 3D printing go down and quality continues to rise, more inventories around the world will inevitably be replaced with on-demand parts, or—the term we prefer—virtual inventory.

In fact, this is exactly what our company, Fast Radius, has created. Direct Digital Manufacturing, as it is often called, allows you to electronically send a design file to our factory and start production automatically, 3D printing as many copies as you need and shipping them to you or anywhere else in the world. In the spring of 2015, we completed construction of the world's first fully automated 3D production factory in Louisville, Kentucky. We built this facility right into the UPS hub facility to create unequaled distribution efficiency, right at the end of the runway.

Amazon's Web Services is referred to as the "Elastic Computing Cloud." It allows users to pay on-demand for computing capacities like processing, speed, apps, and storage without having to buy a computer. Our facility is the first hub in what will become known as the Elastic Manufacturing Cloud, a giant step in the future of supply chain and manufacturing. It allows you to manufacture and distribute at scale with nothing more than a digital file. Need parts in Iceland or Uruguay but want to avoid month-long delays in customs? From wherever you are, simply go to the cloud and order what you need, where you need it, in the exact quantities required. Need one million parts produced in two days? Tap into the global system, which will automatically find you one million printers in the locations closest to you.

The fact that our 3D micro-factory was installed in a UPS facility is more than just symbolic. We are partnering with the company as part of their effort to stay ahead of the curve. As the economy shifted to a new Internet-dominated environment, UPS realized that they needed a fully distributed and networked system of distribution. Now, 3D printing is just beginning to shift manufacturing. To continue leading, UPS is beginning to offer 3D printing capacities at its stores around the country. They are experimenting with a 3D printed distributed-production model—perfectly overlaid on their existing network. For UPS, 3D printing is a truly transformative new supply chain technology.

A CLOUD IN SPACE

In 2015, a custom tool was designed on Earth and sent to the international space station. Astronauts downloaded the file and printed it out on their onboard 3D printer. This is maybe the most mind-bendingly sci-fi use of 3D technology to date. NASA *emailed a tool to outer space* as easily as you might text a friend. Theoretically, they could have sent it anywhere that receives or retains data. Automated 3D printers could one day build and repair settlements on Mars with continually updated designs from engineers on Earth. It's a stunning demonstration of how 3D printing rips reality, reducing the physical world to files. And it makes non-digital manufacturing look like snail mail.

In fact, a lot of what initially seem to be primarily technical breakthroughs in manufacturing will disrupt the global supply system. For example, GE's fuel nozzle that reduces twenty-one parts to one shows 3D printing's unequaled capacity to print complex, previously impossible objects. But it will also dramatically affect the supply chain. The fewer parts that need to be assembled in a centralized location, the fewer needed to be delivered there. In the case of GE's nozzle, that's 20 parts multiplied by the 400 units currently on order: 8,000 parts taken out of the supply chain. That huge reduction comes from just one relatively low-volume component.

Another possible disruption is consumers printing consumer parts on home printers instead of having them delivered. This is probably the most overhyped dimension of 3D printing, but improvements in hardware, design, and materials are rapidly continuing. Depending on how desktop 3D printing technology develops, home printing could create absolute chaos for logistics services.

This disruption to supply chain companies was not expected during 3D printing's early days. When the technology was just used for rapid prototyping, no one thought it would have a major impact on trucking or shipping. But as 3D printing has moved into end-use parts, a technology that was once seen as a better way to build things has emerged as a menace for logistics companies. According to estimates by global consulting firm PricewaterhouseCoopers, 3D printing will be so disruptive that up to 40 percent of traditional air and shipping cargo will eventually be under threat. Thousands of companies who participate in this global system of production and distribution will ultimately risk becoming obsolete—unless they change their business model.

GLOBAL SHUFFLE

American Holt is a Massachusetts-based company that supplies after-market parts to food, beverage, and consumer goods companies. Today, most of their business is organized as a traditional supply chain. When a component in one of the machines that the company services breaks, American Holt often reverse-engineers the part in-house. Then they send the design off to be manufactured by one of their partners. The manufactured parts—usually small components like gears—are then shipped back to American Holt, which warehouses them until they are needed. This technique, creating spare parts with several years supply and then warehousing them, is a key anchor of global supply chain. But many parts are only needed in limited quantities, producers have minimum order quantities in the hundreds or thousands, and holding costs are estimated at 25 percent of the cost of the part every single year. Inventory is value in purgatory. International logistics present other problems for many companies.

When machines that American Holt services abroad break down, the company can't always rely on the local infrastructure. Domestically, American Holt aims to deliver parts within one or two days. But shipping parts to some countries in Asia, Africa, and South America can take a week, during which time critical machines aren't operating. Even more importantly, the imported parts could rack up import taxes and agonizing customs delays. Some countries collect up to 300 percent tariffs, and can delay shipments by as much as two or three months in customs. The alternative is now to print in country. Today, American Holt is expanding its 3D printing capacities and exploring partnerships with printers locally in developing countries.

Many other companies have the opposite problem. To ensure timely access, they warehouse replacement parts in countries around the world. But, eventually, it no longer makes sense to hold some of the inventory in a particular country. It may be outdated or in demand somewhere else. This is a problem. Due to national regulations in many countries, it is incredibly difficult to get the inventory back out. By installing 3D printers locally, these companies can produce parts on-demand and avoid import fees and trapped inventory.

There is also huge value in local, 3D production for consumers. Today in Saudi Arabia, for example, if you dent the fender on your Toyota Camry, you have to order the replacement part from a dealer outside of the country. It takes weeks to arrive and is subject to very high import duties. The same part printed locally could be ready in days and, because it is manufactured locally, avoid duties.

These are real problems faced by real companies every day. Cloud-based Direct Digital Manufacturing offers an alternative. Locally based 3D printers allow companies to reduce warehousing with virtual inventory. They can print on-demand how much you need, exactly when you need it, and in whatever location.

A hundred years ago, the car was rapidly adopted in large part because the train could only go back and forth on tracks. This is the sort of inflexibility associated with supply chains today. As

Second Wave technologies like 3D printing and the Internet converge, a less linear network will emerge. In fact, it will look a lot like a social network.

THE NETWORKED COOKIE

Is it possible to make a cream-filled cookie part of a digital social network?

In 2013, this was the question that Twitter executives posed to Mondelez, the company that makes Oreo cookies. Various managers from the Oreo division were intrigued. Twitter was even willing to throw some money into a project to create a networked vending machine for Oreos. But there was a catch. Twitter wanted to premier the machine at the South by Southwest (SXSW) festival in Austin, Texas—in six weeks. That meant a small team had to design, prototype, and build a machine capable of 3D printing tens of thousands of customized cookies in a day. The machine had to be approved by Oreo food scientists and Austin health authorities. Oh, and it had to look cool.

Steven Spencer, a senior designer, became a lead on the loosely conceived and ridiculously urgent project. Spencer started by brainstorming with his colleagues. His engineers created Computer-Aided Design files of how the machine might work. For the front of the machine, they wanted a clear multitouch panel so that people could order their cookie and then see it being made. They ordered high-definition displays that could change from transparent to completely opaque directly from secret Samsung labs in Korea. Behind the display, customers would see an arm swing over with a selected flavor of wafer. A modified 3D printer using surgically safe medical dispensers extruded any combination of twelve different flavors of icing on the wafer. Then a second arm would swing around with the second wafer. The cookie was dropped in a paper cup that shot down the vending slot to the customer.

In another room, about a dozen people were staring at computers trying to figure out how the cookie was going to interact with Twitter's social network. The project name was "Eat the Tweet." Eventually, the team decided that cookies could be selected based on topics that were trending on Twitter. Customers would be able to pick different flavor combinations of icing filling and wafers based on what was trending. If they wanted to further customize, they could do mash-ups, combining the cookies associated with two different trends. So far so good. But, still, nobody had ever made a tweet edible. This was uncharted territory. If Kim Kardashian were trending, what would her cookie taste like?

After six weeks of nonstop experimentation, two "trending vending" machines were flown to Austin in a shipping container. The container was set up on a street in Austin. Next door was an "Oreo lounge" with seats and milk. Then customers showed up—in droves.

During the nine days of the South by Southwest festival, over ten thousand people waited in line to Eat a Tweet for an average of two hours. The vending machine also generated 45 million media impressions. The socially networked cookie worked! The lines stretching around the block were all the proof that was needed.

As exciting as it was, "Tending Vending" was ultimately about making a fun cookie for a festival. What did it mean in the larger world? For one, it showed the appeal of local, real-time customized production. If we can make a socially networked cookie, then why not shoes or industrial parts dictated by local trends? This is a step beyond hyper-local manufacturing, it's allowing people's wants and interests to actually influence what is produced. It also shows how consumers, producers, and designers all are becoming part of a broader, connected market. Products are not forced upon you. Everyone becomes part of the design cycle in real time. Manufacturers and supply chains and consumers are all becoming part of a great big interconnected social network.

Simultaneously, the Internet of Things is placing products in their own social network. Parts interact with other parts. Products that show wear much more quickly than expected generate immediate attention just like an idea that is trending on the Internet. In this world, there's no place for time-consuming, clumsy, and linear supply chains.

WHY 3D PRINTING WON'T KILL THE MAILMAN

In some ways, 3D printing's unexpected but inevitable transformation of supply chain is reminiscent of how the printing press undermined the Catholic Church. The church was an early adopter of Gutenberg's printing press, initially viewing the machine as simply a better way to print church literature. But after just a few decades, the church recognized the technology's threat to its monopoly on information. In the 1500s, three different popes banned the unauthorized sale or manufacture of books. But it was too late.

Today, anyone who sees 3D printing as just "a different way to make things" is in for a similar shock. No, 3D printing isn't going to be the "end of shipping," any more than the printing press was the last gasp of the Catholic Church. What is certain, though, is that the supply chain in a 3D printed world will look very different. The challenge is to know where the changes are most likely to occur, and race quickly to position your organization for that future. As Indy race car driver Mario Andretti said, "If everything seems under control, you're not going fast enough."

7

THE FUTURE IN 3D

I confess that in 1901 I said to my brother Orville that man would not fly for fifty years. Two years later we ourselves made flights.

—WILBUR WRIGHT

IN EARLY JUNE OF 2010, a twenty-eight-year-old named Khaled Saeed walked nervously into a cyber-café in Alexandria, Egypt. He had recently posted a brief video clip of Egyptian police officers dealing drugs. His friends and family had warned him to be careful. The Egyptian authorities did not respond kindly to public criticism, much less incriminating evidence. Saeed sat down at a computer in a corner of the second floor. Moments later, two policemen stormed in, grabbed Saeed, and dragged him across the street. There, they repeatedly bashed his head against an iron door, killing him. Saeed fell onto the street, where the police continued kicking him.

Cut to six months later in a small provincial Tunisian town in Northern Africa. Police have confiscated the fruit of a twenty-six-year-old vendor named Mohammed Bouazizi, who had been unable to pay a bribe. Emotional and frustrated with the lack of political freedom and economic possibilities in his country, Bouazizi ran to a gas station and filled up a canister. He then walked into a busy intersection and doused himself with the gas. While yelling "How do you expect me to make a living?" he set himself on fire.

These were both very public and gruesome deaths of young men. They pointed accusatory fingers at their nations' political structures. But to anyone familiar with the history of Egypt or Tunisia, the despondent response may have been: "It won't change anything."

Egypt had been ruled by dictators for decades. Then-President Hosni Mubarak had been in power for thirty years; his tenure had been full of human rights abuses. Mubarak had survived numerous assassination attempts by political opponents. He had successfully repressed a well-organized and popular opposition group, the Muslim Brotherhood, by arresting their leadership. In jail they were routinely tortured and even died. What difference did one more unjustified death make?

Tunisia's President Zine El Abidine Ben Ali had also ruled for decades without competition. While his regime was not known for the same level of repression as Mubarak's, Tunisians implicitly understood that there would be no substantive government accountability for Bouazizi's fatal protest.

But something had changed in the world. During a visit to the morgue, Saeed's brother snapped photos with his cell phone of his brother's mangled face. He posted them online where they quickly went viral, provoking national and international outrage.

A few days after Bouazizi's self-immolation, people poured into the streets in protest. The subsequent police crackdown was filmed by residents, who posted videos online on YouTube and other sites. Similar protests followed. After decades of power-lessness in the face of oppression, the Internet had made political protest and organizing a completely different game.

Social media allowed people to communicate around the coun-try and world much more easily. Twitter and Facebook had origi-nally been intended for networking among friends, but they were now hijacked for new purposes. The Facebook page "We are all Khaled Saeed" became a virtual rallying point for discontented Egyptians. The Twitter hashtag named for Mohammed Bouazizi's hometown, #SidiBouzid, served a similar use for Tunisians. The results of the protests were swift and stunning. Less than two months after Bouazizi lit himself on fire, President Zine al Abi-dine ben Ali fled into exile—after twenty years of rule. A month later, Egypt's Hosni Mubarak stepped down and was arrested for human rights abuses.

In just months, Egyptian citizen activists accomplished what the established, centrally organized Muslim Brotherhood had failed to do for decades: overthrow Mubarak. The media widely reported that Twitter and Facebook's role in these revolutions was spreading information swiftly to a large number of Egyptians. But this is an incomplete assessment. In the pre-Internet era, millions of Egyptians were well aware of the illegal arrests, political sup-pression, and police beatings carried out by the government. De-scriptions of state crimes spread by word of mouth throughout these densely populated communities with or without social media technology. What Twitter and Facebook enabled wasn't access to information, but the shape of the protest that emerged.

In the past, Mubarak would squash political protest using the tried-and-true technique of dictators everywhere: swiftly arresting the leadership, perhaps torturing or killing them to boot. But this time around in Egypt—as in Tunisia—there was no one to arrest. Twitter, Facebook, and other social media encouraged a non-hierarchical organizational structure. Identities of Internet activists were harder to detect and a single voice quickly blended in with millions of others. Even if you arrested a few dozen people, it was easy for others to step in and continue using the decentralized and open channels of communication. The authorities would just be playing whack-a-mole while enraging the growing crowd.

Another common media commentary was how surprising it was that these protests had gained strength without a charismatic leader. In fact, it was precisely *because there was no visible leader* that they succeeded where others had failed. People who had never been engaged politically before—either out of cynicism or fear—suddenly felt safe jumping in, and they joined the protests by the millions. Twitter and Facebook, for the first time ever, enabled a "leaderless revolution," and these tools have forever increased the power of oppressed people around the world.

Facebook originated as a dating site for college students. None of its founders could have imagined that it would be an infrastructure used to overthrow dictatorships. No manager at Twitter ever suggested, "Hey, let's create a revolution app!" But once the companies had put these new technologies into the world, there was simply no way to predict how people would use the new tools. That's the amazing thing about Second Wave technologies like web-enabled social media. They empower ordinary people to make modifications and find new-use cases—not just the website and app creators and owners. First Wave technologies like the train result in huge social and political changes, but they don't tend to encourage people to use the new technologies for their own purposes. In fact, some of the most important social and political innovations of previous centuries, including the labor movements, were formed in direct opposition to First Wave practices.

Second Wave technologies are not oppositional in this sense. Instead they encourage redesign. This promotes new thinking, but it also makes it much harder to guess which way innovation is going. When people started using the Internet to send emails in the early 1990s, not a single record executive imagined that pirated downloads would change the nature of their business more than the invention of compact discs.

The same is true of 3D printing. As the technology's adoption spreads, its use cases will explode. Some of these outcomes are relatively safe bets. For example, it seems extremely likely that medical and aerospace companies will find new efficiencies in customized and complex 3D printed parts. But the innovations of a few years from now are much harder to predict. All we can be certain about is that 3D printing's lowered barriers to access will cause manufacturing innovation to explode. We can't predict the future. We just know that the tech will be used in ways that no one could have predicted, to produce outcomes that no one could have ever imagined.

MAKING A BIGGER WELCOME MAT

Noah Fram-Schwartz, the twenty-year-old founder of a Silicon Valley start-up called Makexyz, was always a builder. Like most kids, he started with Legos and moved on to tree houses. But by high school, he was on to much more complicated projects.

"I used to do a lot of high-magnification photography as a teenager. I built this huge rig, took apart my microscope, and mounted my camera on it. I had a bunch of X, Y, Z slides that would allow me to move the camera in increments of half of a micron. I had to machine a lot of the components myself. I remember cutting out a corner of an aluminum part with a crappy hacksaw and then trying to drill an angle. It took me a day to make one part."

Fram-Schwartz wasn't discouraged by the difficulties of being an amateur inventor. But when he first saw 3D printing, a bell went off. "I thought, 'This is it!' You can go from the idea in your mind to the idea in your hand—sometimes in minutes if it's small enough."

By the time he enrolled at Brandeis University, Fram-Schwartz had moved past desktop 3D printing and on to larger, professional machines. He began running a small service bureau in his dorm room, manufacturing 3D parts for clients, including medical companies. An obsessive worker, Fram-Schwartz studied computer science, product design, and Chinese while starting Deis3D, Brandeis' 3D printing club. In March of 2015, Deis3D hosted teams of students from the University of Connecticut, MIT, and Columbia University for the world's first 24-hour Print-a-thon, modeled after the software "hack-a-thons" common in Silicon Valley.

Each team came in with a design based around the theme of "social justice." The winning team built a prosthetic leg that attaches to a bicycle pedal, allowing amputees to ride bikes. Then, in the middle of his sophomore year, a job offer came in from Google; Fram-Schwartz dropped out of Brandeis and moved to California.

"I went from being a college student living off of ramen to being tasked out to these labs with these enormous budgets," said Fram-Schwartz. "I learned a lot. I tried to talk to everyone. And I spent fifteen to twenty hours a day on hardcore research about 3D printing." Then, after just over six months at Google, Fram-Schwartz left behind the financial resources of the massive tech company with the same conviction as previous generations of Silicon Valley Internet entrepreneurs. Fram-Schwartz knew that Second Wave technologies flatten the playing field. He knew he didn't need to have unlimited financial resources to launch his own start-up, Makexyz. In the Second Wave world, the most important competitive edge is not established market position or value, but innovation.

In fact, Fram-Schwartz's start-up doubles up on that concept of lowered barriers. Professional engineers are able to optimize 3D designs with software like Autodesk Within. For an amateur, this represents a substantial investment. "You usually need a trained, certified engineer to spend a long time playing with the different parameters and specifying the loads and all these other constraints," said Fram-Schwartz. Entrepreneurs chasing manufacturing ideas or service bureaus that handle one hundred to two hundred orders a day simply don't have a way to fully explore 3D's possibilities.

So Fram-Schwartz decided to try and lower barriers to 3D printing's innovation by automating a portion of the design process. Specifically, he is creating technology that seeks to maximize 3D printing's unique ability to produce objects with a lattice core. We talked before about how lattice structures provide huge mechanical advantages. They reduce weight while maintaining strength and performance, enabling lower material and production costs, and encouraging design customization. In one example, a researcher at Purdue University used his spare time to print a small 3D latticed cube weighing 3.9 grams (.008 pounds) and performed a crush test on it. The cube was able to support

900 pounds—or over 100,000 times its weight—before collapsing. So why isn't everything manufactured with a lattice core? First, they can be built only using 3D printing. And, second, even if you have access to a printer, you'd still need some serious brainpower to design the lattice.

Fram-Schwartz offers an alternative. Instead of paying an engineer, you use an algorithm that unlocks customized complexity for users who wouldn't otherwise have the necessary financial resources. By automatically regenerating the designs created by nonexperts using lattice structures, his algorithm is able to optimize one of 3D printing's most powerful capacities for complex geometries. By lowering the entry barriers to 3D innovation, Fram-Schwartz's application also encourages more widespread and better use of 3D printing.

Noah Fram-Schwartz is just one of a million young entrepreneurs who are hoping to radically change a market. It is entirely too early to know if his new company will ever gain enough traction to survive, let alone flourish. But what is important is that Fram-Schwartz is motivated to be a manufacturing entrepreneur. The Internet has lured millions of bright young people away from traditional career paths with the hope of creating the newest technology or the hottest app. Now, motivated by the sweeping new opportunities that 3D printing is creating, waves of entrepreneurs are turning their attention to production innovation for the first time in decades.

3DSIM is another entrepreneurial company hoping to lower the barriers to optimized 3D printing. Aimed at companies that can afford metal powder printers but don't have in-house expertise to optimize them, 3DSIM improves printing at the molecular level. The thermodynamic process inside the machine creates a series of incredibly complex engineering questions. 3DSIM [tagline: "Because you don't have 5.7×10^{18} years to wait for an answer"] optimizes machinery in ways that will produce parts with fewer distortions. Of course, at around a million dollars a pop, metal printers present a substantial cost for the average college dropout. But another company, Desktop Metal, is working on smaller, cheaper printers that print metal without the use of lasers.

Each of these companies is playing a role in encouraging manufacturing innovation. First, they have their own innovative products. But, more importantly, they are further lowering the barriers to entry to 3D printing's full capabilities. The rate of innovation is directly proportional to the number of participants. 3D printing technologies are bringing legions of new people into the circle. Manufacturing is becoming cool again.

In other words, some innovation will occur because a flexible new technology is available. Much more innovation will happen when that technology also spurs greater access and competition. We are now seeing innovation across the board, from start-ups, entrepreneurs, and amateurs posting designs online, to large companies seeking to maintain their innovative edge. These tools, combined with the elastic manufacturing cloud that we discussed in chapter 6, will unleash an entirely new world, one that doesn't look much like the one we live in today.

3D PRINTING GOES BIG

Most 3D printing has been limited to the size of the build plate or machine in which it is printed. But there are already several people and companies breaking through that barrier. Joris Laarman, for example, is the founder of a Dutch start-up called MX3D that has developed a 3D printer capable of creating plastic and

metal structures in midair with no printer bed. To advertise the technology, in the fall of 2017 Joris will bring a printer to each side of a canal in Amsterdam and turn them on. The printers move like a train that creates its own tracks as it rolls along. Over the following two months, with no human guidance or scaffolding, the machines will use these tracks while printing a functional, ornate footbridge over the water. The machines will meet in the middle. If successful, the other uses for what is known as "free-form construction" are almost limitless. If huge buildings could be built cost-effectively without scaffoldings or manpower, the construction industry would undergo a process of massive transformation.

In fact, houses and apartment buildings have already been built by giant printers. In China, printers that extrude a cement-like mixture are mounted on huge robotic arms. They race around a construction site, building walls from the ground up as effortlessly as a baker squirts icing on a cake. In early 2015, a company called WinSun claimed to have printed ten houses in just one day. The technology is one idea for rapidly providing housing for the country's and world's exploding population. Some printers are able to use more sustainable and local construction materials. WinSun's printer uses a mixture that includes waste as well as cement and other construction materials.

In Italy, a construction company called WASP built a forty-foot tall twenty-foot wide printer that can use clay, dirt, and other common local materials. Their idea is to create a machine that can build structures in poor areas where the infrastructure needed to transport traditional building materials is limited.

Designs sent from earth have already been manufactured by printers in orbit. Maybe NASA will launch an exclusive "Made in Space" brand? One company is planning on inverting the build envelope to avoid the constraints of a cramped workspace. The machine would 3D print outside of the spacecraft, giving it literally limitless space.

Printers in space also have a unique value. They can be used to replace and repair pieces of equipment without the painfully long and expensive supply chain issues of rocket-based delivery. The potential for this printing could have a huge impact on future civilization. NASA is developing printers that can build shelters from the grit that blows across Mars. Meanwhile, the European Space Agency hopes to print an entire moon colony with its already existing machine that 3D prints 1.5-ton building blocks using lunar dust.

3D PRINTING GOES SMALL

On the other end of the spectrum are tiny 3D printed lithium-ion batteries. Developed by a team at Harvard led by a materials scientist named Jennifer Lewis, these fully functional power sources are as small as a millimeter square, but work as well as normal batteries. This is possible because Lewis has used extremely accurate syringes and newly functional inks to work on normal 3D printing extrusion machines—the same idea Scott Crump used to print a frog in his kitchen in the late 1980s! These batteries, in turn, could be used to provide power for microscopic self-powered biomedical sensors that continually transmit data to a smartphone. They could power hearing aids so that batteries didn't have to be replaced. But, most importantly, they open up yet unimagined pathways for other novel devices.

Computer hardware companies have continually made their silicon chips thinner and thinner. Today, Intel's processor chips have gotten down to an infinitesimal fourteen nanometers thick or about the radius of 6 strands of DNA. But at that size, they are approaching a fundamental limit—the atomic radius of silicon. Soon silicon may be replaced as the primary material for the chips that power our computers. One possibility is electronics printed out of plastic. With 3D printing, they could become smaller at an exponential rate. Some technologists think that could lead to plastic circuits with performance levels equal to today's silicon chips—but at a fraction of the cost.

Another option to replace silicon is graphene, the 2D super material. Because the carbon atoms that make up graphene have a smaller radius than silicon, the chips would be thinner. A 3D printed nano shape could serve as a mold—or as the circuit itself—of graphene. But because of graphene's other amazing properties, the chips would also be much more heat resistant. Graphene's use in other printed electronics could also be disruptive. In a radio, they can handle frequencies much higher than silicon versions.

Graphene routers would make today's wireless speeds look like a jalopy on a freeway.

3D PRINTING COMES ALIVE!

It used to be there was just one way to get a steak—or a pair of leather gloves. But now a company called Modern Meadow is 3D printing beef and leather in their lab in Brooklyn without harming a single animal. Biofabrication takes cells from animals through a noninvasive biopsy. Who knows if 3D printed beef will have a following. But we are now able to print a square foot of leather just as fast as raising it the old-fashioned way. That doesn't just save a cow's life, it saves tons of feedstock and eliminates manure and methane. For now, however, leather and printed hamburgers are just the headline grabbers in a much more meaningful convergence between biology and 3D printing.

In 2002, a Japanese scientist named Makoto Nakamura realized that the droplets of ink expelled by his home printer were roughly the same size as human cells. Intrigued, he tried to print out a biological object. His first attempt was a failure. The biological matter clogged the nozzle on his ink-jet Epson. Nakamura called Epson customer service for help. Not surprisingly, his requests were turned down several times. Most associates didn't think that their job included explaining how to print human tissue. Finally Nakamura got a manager at the company to give him the technical support he needed to adjust the nozzle. The next year Nakamura was printing out living cells in a lab on a modified home printer. Since these home brew beginnings, the medical use of 3D printing technology has exploded.

We already saw doctors scanning knees to print personalized

replacements. The next level is producing kneecaps in which the print material itself actually contains antibiotics and pain medication. Both of the medicines are slow released after surgery over nine months, eliminating the need for some follow-up visits and pills and dramatically reducing the risk of infection.

A San Francisco–based company called Cambrian Genomics recently unveiled a 3D laser printer that can print DNA ten thousand times faster than traditional methods. DNA is the software of life, cells are the hardware—and everything living is made from those building blocks. While we've been able to produce DNA for years, it was incredibly slow and expensive. Cambrian Genomics' new technique can print more DNA in a single run than all the other machines in the world print in a year, dramatically reducing costs. Cambrian claims that the technology will eventually be cheap enough that DNA—the code of life—will become a consumer product. The possibilities of customizing life are endless and possibly frightening. They could be used to create artificial life or, the company says, to print out living versions of extinct species.

A use we're more likely to see in the near future will be writing code that attacks pathogens like cancer. By hijacking virus cells, the DNA could be used to insert a unique gene into a patient's DNA. The technology could also be used outside human bodies by creating tissue of, say, mammary glands, out of the DNA of a breast cancer patient. The tissue could then be used to test potential treatments before giving them to the patient.

In 2014, printed body parts like hips and knee replacements were worth $537 million in the United States, a 30 percent increase over 2013. The next frontier is printing biological replacements. A company called Organovo prints liver tissue for pharmaceutical companies to do toxicity tests without harming living creatures. The next goal is printing patches that can be used to replace or repair damaged liver tissue in humans. Eventually they hope to print entire organs for transplant.

3D PRINTING ENTIRE OBJECTS

The earliest 3D printers manufactured in plastics, one reason they were useful mostly for novelties and prototyping. Since then, the number of materials available for printing has expanded rapidly to include, for example, ceramics, metals, genetic matter, food, and various alloys and composites. Still, most 3D printers today work best with a single material or, at most, three of them. But that, too, is changing rapidly. Stratasys has a model that can print in forty-six different materials or colors. Recently MIT's Computer Science and Artificial Intelligence Laboratory has introduced a much more affordable machine capable of using up to ten materials.

Advanced and fully functional multi-material printing is considered the Holy Grail of 3D printing by many experts. It would allow machines to print entire consumer electronics in one piece. Although this technology would create manufacturing efficiencies—no assembly line required—it's unlikely these products would ever compete with mass-produced electronics. But they do make possible assembly-free and completely customized electronics. The technology is also crucial to the development of the Internet of Things. The advantages of seamlessly embedding multiple sensors in products during production are huge. Otherwise "dumb" pieces of plastic or metal—say, load-bearing structures—could become smart by communicating to other devices and people. Another future application of multi-material printing: embedding actuators and motors into robots or drones that could walk or fly off the build plate.

Disney's research division is also working on integrated components, including what are known as light tubes. Like fiber-optic cables, these tiny passageways can direct light from point to point through an object in nonlinear directions. But light pipes have two distinct advantages over conventional manufacturing. First, they can be printed in one pass with customized widths or connections with other pipes. Second, light pipes reduce the number of individual parts and assembly time.

One of the researchers' proof-of-concept examples was pure Disney. A cartoon toy figure was placed on a lighted platform and the light was directed through the feet and then upward and forward to light up the eyes. But this was a very smart toy. The eyes became display surfaces, responding to user interaction such as sound or movement.

Miniature light pipes were also printed into chess pieces. When players explored moves, the light detected and gauged movement. The pipes also turned the back of the piece into a display that suggested moves. By themselves, these early applications are hardly earth shattering. But the potential for embedding custom optical elements that can display information and sense user input in virtually any object? You could turn a fridge into a keyboard or a wall into a monitor.

Another kind of 3D enabled integration came to life during a public-private partnership that explored merging the energy usage of cars and houses. During the Oak Ridge National Labs' Additive Manufacturing Integrated Energy project, teams prototyped and 3D built a novel electric vehicle. Then they used twenty-five thousand pounds of 3D printed material to build a house out of arched segments. The idea was to architect a house with zero design constraints including corners and straight walls. The result looked a bit like a futuristic whale's rib cage covered with solar panels.

The most interesting development was the symbiotic relationship between the 3D printed car and house. When you drive down an interstate at sixty-five mph, you are producing enough energy to power two of your houses. So when a driver gets to the 3D printed home, she connects the car to her house via a bio-directional charger. Depending on the time of day and other factors, the car's battery may help power the efficient house or energy from the solar panel, or wind sources may be stored in the car's battery.

4D: COMING SOON TO A PRINTER NEAR YOU

If the potential for 3D printing isn't enough, imagine printing objects capable of rearranging themselves. Computronium, or programmable matter, is a fictional substance that appears in science fiction from *The Hitchhiker's Guide to the Galaxy* to *Iron Man* comics. It is also imagined as a smart material that changes its physical properties, such as shape, weight, density, and conductivity, at some point in the future based on sensing external input. Now it's real.

Think of how a dry towel expands and changes shape when water is added to it. A 4D printed object isn't limited to its original shape. In fact, it is designed not to retain its original shape. Led by Skylar Tibbits, MIT's Self-Assembly Lab is printing objects in metal, carbon fiber, and wood that rearrange themselves in response to various environmental changes, like light or temperature. In one remarkable demonstration, a metal chain laid out straight on a surface automatically bent itself until it spelled out "MIT" once it came in contact with water. The chain did not use motors or any control devices and was not attached to any power source. The behavior was internalized in the build material. Just like the metal chain, a wide range of materials can be programmed to change or carry out an action in a desired way at some point in the future.

Rubber can be 4D printed—and eventually used to change tire density based on road conditions. Work with wood grains may produce flat shipments that assume their final shape, like chairs, after the package is opened. Other 3D printed materials have been programmed to expand up to 200 percent of their original volume. Imagine a brick that does not realize its final shape, structure, or weight until water is added to it, at the exact location that it will be used.

Researchers in Australia have printed a valve that automatically closes itself when the water surrounding it reaches a certain temperature. The valve doesn't need any activation; it comes out of the printer ready to use. This technology could easily be adapted to detect wetness and build pipes that repair themselves during a leak—or self-changing diapers. But the fact that 4D printed materials do not rely on electricity means that they can create previously impossible life-saving applications. In medicine, for example, new electronic devices that don't need power supplies could be implanted.

Printed parts could also communicate with each other to improve performance. For example, when a car is making a sharp turn and there is stress on one side, an electronic signal could be sent to the material itself, making it stiffen up.

Other 3D printed applications include what is known as shape memory alloy, where a change in light triggers a shape change. The same reactions can be caused by electrical and chemical stimulus. Shoes could change functionally when it begins to rain, or a path turns rocky. Jackets will change density or wind reflection when it gets cold or windy. The military is developing chameleon-like 4D printed camouflage that actively blends into the surrounding environment.

Airbus wants to reduce drag caused by a robotic piece that regulates air to the engine. It is working on a 4D printed piece of carbon fiber that flexes in and out depending on engine heat. Because the 4D piece doesn't need motors, sensors, or even electricity, it will eliminate size and maintenance problems.

Just a few months ago, BMW announced details of a concept car ripped straight from the future. The Vision Next 100 will be 4D printed, with every part functioning right off the print bed. BMW envisions a future where additive manufacturing machines will combine all necessary raw materials—silicon, metal, and more—into the end product in a single print. Among other proposed revolutionary improvements, The Vision Next 100 is equipped with an outer shell that flexes and changes shape based on airflow and steering movement.

NEW WAYS TO PRINT

In this superheated mode of innovation, it's hardly surprising to see new technologies inspired by fiction. In 2013, a group of scientists and entrepreneurs decided that they were going to develop a printing technology that specifically mimicked *Terminator 2*'s replicator villain as it rose out of a liquid vat. The process, called Continuous Liquid Interface Production, or CLIP, is actually similar to stereolithography, one of the earliest 3D printing technologies. But instead of pausing between increments, the resin continuously moves through a UV light. As a result, there is no layering effect in the final product. The object rises up continuously as it is being created. The technology has been greeted with breathless enthusiasm. Investors including Google Ventures announced an investment in 2015 of over $100 million in Carbon3D's new technology, which claims to be up to one hundred times faster than current 3D printing.

Just as Carbon 3D was announcing its new technology to the world, an engineer in Canada named Diego Seoane was furiously working to design a new printer that he hoped would double or triple the speed of a 3D printer. When the media began heaping praise on Carbon 3D's new accomplishment, his plans collapsed. Seoane's backers pulled out their investment in his research. Down but not out, Diego went back to the drawing board. Within six months he had uncovered another 3D printing breakthrough. In January of 2016 Diego unveiled his new 3D printing technology at the Consumer Electronics Show. The machine prints an object right before your eyes—at speeds up to two times faster than a Carbon 3D printer.

If either of these new technologies lives up to their promise, they will join the cell phone as another sci-fi inspired technology that changed the world. Then again, why shouldn't sci-fi be the muse of this new technology? The former depends on limitless imagination; the latter, on limitless design freedom.

In 2014, HP announced that it had developed an industrial 3D printer that is ten times faster and 50 percent less expensive than other machines. The Multi Jet Fusion works by using what looks like a scanning bar on one of HP's typical 2D printers. In fact, it is a 3D print bar embedded with thirty thousand nozzles that spray 350 million drops of material a second as it moves back and forth across a print platform.

The speed and sophistication of these machines is very impressive. But it is the printer's capacity for extreme customization that is most exciting. Today, nearly all 3D printers create objects in layers, using the same materials throughout. A 3D printed plastic part has the same properties of material in every layer.

In the new HP technology, the printers add an extra step. The printer first spreads a micro-thin layer of material on the build platform. Then the print bar adds a transforming agent, which can change individual properties of the material. For example, the printer can control for color, texture, friction, strength, elasticity, electrical properties, thermal properties, and more. But what is truly astounding is that these agents can be customized to areas as small as 1/10,000 of an inch! Imagine a single part, with stiffness optimized in some areas, elasticity in others. Or wear resistance and friction customized exactly where needed. Or imagine printing a complete electromechanical module in a single 3D build, without requiring any further assembly. The list of future possibilities seems endless.

Look at every object around you. How were they manufactured? What materials were they made from? Now imagine them consisting of millions or billions of customizable points that together form the physical object. This new printer holds the potential to produce objects unlike anything we have ever imagined.

3D printing remains a disarmingly simple technology: You can print anything in layers. But its simplicity belies vast possibilities for innovation. Today, 3D printing is in an explosive growth period, a spiral of innovation and investment that will spur development of new technologies and materials. The new technologies will uncover new applications, and these new applications will spark even greater technology development. It's impossible to accurately predict exactly what will be happening in five years, much less ten or fifteen. But with low barriers to entry and the $14 trillion manufacturing industry now in play, it's game on.

8

SYNCHRONICITY

Culture does not change because we desire to
change it. Culture changes when the organization is
transformed; the culture reflects the realities of people
working together every day.

—FRANCES HESSELBEIN

BACK IN 2011, GENERAL ELECTRIC was a company behind schedule. CEO Jeffrey Immelt had enunciated a new corporate strategy a decade previously. He envisioned a more nimble and entrepreneurial company that focused on technology and infrastructure. His direction was starkly different from that of his legendary predecessor, Jack Welch, but he was dedicated to carrying it out, despite daunting odds.

Four days after Immelt was named CEO, terrorists flew two

commercial airliners into the World Trade Center in Manhattan, just an hour south of GE's corporate headquarters. It was a sign of things to come. For the rest of the decade, Immelt found himself putting out fires. Ten years after he had taken the reins, Immelt's vision was still unrealized. GE employed thousands of talented engineers and designers, but tech companies were still more innovative. With the iPhone and iPad, Apple had owned the 2000s. In 2010, Apple's exploding stock prices drove it past GE to number five on the list of largest companies by market capitalization.

But the financial success of massive companies like Oracle, Cisco, Apple, Google, and Microsoft wasn't what intrigued GE executives. They were more fascinated by the creativity unleashed by tiny tech start-ups. Though many of them crashed and burned, these smaller companies seemed driven by a feverish quest for innovation reminiscent of General Electric's founder, Thomas Edison. By comparison, industrial manufacturing—even the modern, highly mechanized sort practiced by GE—looked inflexible and clunky. GE decided to see what they could learn from these start-ups.

In 2011 the company hired a Cisco executive named Bill Ruh and charged him with opening a software research center in San Ramon, near Silicon Valley. The facility gave GE a presence and ability to recruit in the tech center of the world, but it had another important function in the company's transformation. Immelt had made growing the industrial Internet that connects GE's industrial machinery with its digital networks a priority. The San Ramon center would provide capacities to process and support enormous amounts of data. This could include maintenance and wind speed information from a sensor implanted in a turbine.

That same year a book called *The Lean Startup* became a favorite of the business press. Written by consultant Eric Ries, the book made the case for a business cycle of rapid innovation. *The Lean Startup* was an exciting blueprint for the direction that Immelt and other GE executives wanted to take the company. But there appeared to be a major mismatch. The book was based on Ries' experiences as an entrepreneur in tech start-ups. These small companies typically had just a few employees and were often focused on only one major product launch. General Electric, by comparison, was a massive conglomerate with all the bureaucracy and hierarchy necessary for managing its operations. No matter how enamored GE executives became of the book's logic—or how many employees read the book—the 120-year-old Dow mainstay simply wasn't wired like a young tech company. The start-up logic of constant innovation wasn't how GE thought or measured things. In fact, it was hard to see how the logic was applicable to any company that manufactured heavy infrastructure equipment.

For example, Ries promoted what he called a Minimum Viable Product. An MVP was the most basic useable version of a product. Ries argued that by releasing an MVP, a company could rush a just-realized concept to the marketplace. The product would inevitably have some bugs in it, but this could turn into an advantage. By using free, crowd-sourced feedback, the company could quickly release an improved version. This was completely different from large-scale manufacturing. Toyota, for example, released just one model of Camry per year. Even that model was likely to include just minor upgrades from the previous year. It cost too much to completely redesign the Camry every year. MVPs were always unfinished and always incorporating new design input. They forced companies into rapid innovation cycles.

True, MVP was an exciting idea—one perfectly suited for the

beta version of an app or operating software. A new version with fixes could be available for free download in just a few weeks or months with essentially no supply chain costs. But GE built wind turbines and locomotives, what the company calls "Big Iron." Customers couldn't download these massive, expensive pieces of metal. More importantly, GE also couldn't afford for their products to have bugs when released. There were huge safety concerns and regulations in many of the industries GE supplied. The process to fix less-than-perfect mass-produced products was extremely expensive and time-consuming. By these start-up standards, GE was sloth-like.

Another lean start-up concept was split testing: offering multiple products to customers at the same time and then making product decisions based on feedback. This practice was fine for software—and might also work for soda or ice-cream flavors. But, again, it was hard to see how it might be applicable to manufacturing locomotive engines.

The Lean Startup ideas were exciting, fresh, and, to many, would have seemed completely incompatible with General Electric's industrial operations. But Immelt was undeterred. Rather than try to implement these processes themselves, General Electric hired Ries to translate his *Lean Startup* business model for the massive industrial manufacturing conglomerate. In 2013, GE introduced FastWorks, a program designed to train thousands of people to become more entrepreneurial. Soon, GE employees were fluent in a leaner language of MVPs and "pivoting"—switching strategies rather than firing leaders when initiatives fail. But leadership quickly found that creating an entrepreneurial culture was very difficult in a large, hierarchical company. For years, GE's improvement had been driven by Jack Welch's Six Sigma approach, a tightly controlled system with an intense focus on controlling costs, quality, and execution. Six Sigma force-ranked employees. It had enormous organizational benefits, but to many, it also made experimentation analogous to failure. The rigid hierarchy dis-incentivized collaboration. And the relentless pursuit of efficiency was counter to the pursuit of the bold and the new.

Instead of changing its corporate culture with FastWorks, GE leaders increasingly found themselves at war with the company's own past.

THE CULTURE TECHNOLOGY

Simultaneously, GE was considering expanding its use of 3D printing technology. This initiative was driven by Christine Furstoss, an engineer and metallurgist by training who had been with the company for decades. GE had been engaged with 3D printing technology for years, initially through relationships with universities around the country, including the University of Texas-El Paso, Milwaukee School of Engineering, and the University of Louisville. In those days, the primary use of 3D printing was still prototyping plastic versions of products before they were built. But as the technology matured, GE began printing metal end-use parts and reducing shipping costs.

Furstoss explains the benefits of 3D printing to GE's supply chain in terms of a part's frequent flyer miles: "A part may start at a foundry in the Midwest. Then we have to go get tubing to connect it in California. Before the next assembly step, we have to polish the inside at a specialized plant in Germany because we can't get inside after we add the next part."

During this multistep process, the company's millions of components accrue tens of thousands of miles traveling around the world. Each step meant another trip that cost money and, even more importantly, time. "We were too damn slow," said Furstoss. 3D printing could shave thousands of miles off a part, saving money on shipping and speeding up turn around. It allowed the company to be faster and more responsive to customers.

3D printing solved many of these problems. The technology could print using complex geometries that were impossible in traditional manufacturing. As a result, what had been multiple pieces—each with its own frequent flyer miles—could be grown as one component in a single location. "We're very proud of our designs," says Furstoss, "but just how many touch points and how many vendors did we want to use? The thought that we can go from twenty-seven parts to three—it's about speed."

By 2011, Furstoss was convinced that 3D printing was an important technology for GE to have in its portfolio. The company was already leading aerospace in its use of additive manufacturing, but it still relied on outside contractors, like Cincinnati-based Morris Technologies, to produce components for use in jet engines or medical implants. The following year, General Electric bought Morris Technologies and transformed the company into GE's center for additive manufacturing. Christine and others thought they were investing in a technology with exciting promises in both manufacturing and materials innovations. But GE's move into 3D printing would have a huge unexpected impact.

As they began integrating Morris Technology into the company, 3D printing continued to impress Furstoss. "When we started to work with the design team," she says, "there were types of performance we could start to imagine and then test and see and then realize that it could be reproducible. That was eye-opening." But she also noticed something about additive manufacturing that had nothing to do with its technological capacities.

Additive's capacity to quickly test ideas through rapid prototyping wasn't just about speed. It also promoted a more collaborative workflow. In fact, it was a way of working that was antithetical to GE's ingrained practices.

Greg Morris, one of the cofounders of Morris Technologies, had noticed a similar impact on his company when they began investing in 3D printing. Greg has a deep understanding of the way people work in traditional manufacturing organizations. After college, he and his brother became the sixth generation to run the family's steel distribution business. The company had been in their family since 1850. Morris has spent years driving around from one midwestern factory to another as a sales rep. He has manufacturing in his blood. Over the years, he saw industrial technology improve and productivity increase—but the organizational structure remained stagnant.

Even by the 1990s, when Morris sold the family business, factories were still organized like they had been in the nineteenth century. Workflow was designed to be completely linear and hierarchical. The production of a part began with the designers and engineers. When a design was finalized, it would be sent off for tooling. Once that process was completed, the design could not be changed—the cost and time involved forbid it. This unyielding process weighed even more heavily on the technicians, machinists, and other people downstream. They were responsible for actually making the imperfect part—and manufacturing it as cheaply as possible.

Even designs that were mechanically sound might be difficult to machine. The engineers might call for levels of precision that were incredibly hard to create in the real world. If a component was un-ergonomically designed, production personnel might have to reach across a part to polish it down. The end result was not just frustrated people on the factory floor, but parts that cost more and slowed down production. The machinists and technicians might be in the same building or company as the designers, but they might as well be in China for all the input they had in the decision.

But once he began investing in 3D printing, Morris began to notice that the technology was having a profound disruptive effect on the hierarchical culture of heavy manufacturing. An improved capacity to produce high-quality prototypes—or even single versions of testable parts—interrupted the linear, top-down progression of manufactured goods. People with different job titles began interacting during the process of design and manufacture. Complex geometries, novel materials, and quicker turnaround gave 3D printing an edge over traditional manufacturing, but the change in workflow was transformative.

"The most powerful benefit of the technology," said Morris, "applied when you were able to get all elements of the process together. You get your ideation people and your designers and engineers along with your technicians and machinists and inspection group so that everyone is collocated and they have a say in how you're going to approach making a part."

The nearly silent 3D printers could be on the same floor as a design team. Instead of sending off designs for prototyping that took weeks, they were manufactured down the hallway in a few hours. Not only could designers and engineers test the end-use quality parts in real-life situations, but machinists and technicians saw huge benefits. In the old industrial factories, technicians and machinists downstream might not get a realistic idea of the object they had to make tens of thousands of before the expensive and lengthy tooling or casting processes were finished. And once they were finalized, the company could no longer afford to make changes in the design. With additive manufacturing, the technicians could have something to hold in their hand and make suggestions about.

It became increasingly apparent to Morris that 3D printing was not just a technological marvel. The new way of making things actually promoted direct engagement between people across all parts of the production process. "When your engineers and designers are just a stone's throw from machines then they can interact with the technicians and ask questions," said Morris. Instead of a formal channel for communication, "it's the off-the-hand 'Hey can you come look at this design?' 'Can you see any problems in how this is going to build?' or 'Hey, Bob, I know you got to manufacture this thing later, how much stock would you add?' or 'Are you going to have issues in post-machining?'"

By design, 3D printing encouraged collaboration between all players. This meant more savings in both money and time. "If you get all the people under the same roof," says Morris, "your chance of having a successful design the first time—or at least first few iterations—is greatly improved as opposed to just throwing something over the wall and saying 'Go make my great design' and they kick it back saying it's going to have all these problems."

REALIGNMENT

As GE's Christine Furstoss began setting up additive manufacturing centers around the country, she came to a similar realization. Additive was forcing people to change the way they approached their work. It was creating new collaborations. It was encouraging experimentation. And people were more invested in their work.

In public, GE heralded 3D printing as the enabling technology for cool new parts like its radically efficient fuel nozzles and minimal-weight, high-strength engine brackets. But the biggest changes driven by 3D printing were occurring internally. The more invested in additive GE became, the larger the organizational benefits to the company. It turned out that this manufacturing technology drove a cultural transformation that lined up nearly exactly with Immelt's FastWorks initiative.

Take, for example, *The Lean Startup* concept of continuous deployment. This theory demands that as soon as code for an application is written, it is immediately sent into production. It wants to get products into the market as quickly as possible to get real feedback. As with many of the book's initiatives, the goal is to reduce innovation cycle time to the shortest possible outcome. On its face, continuous deployment seems completely incompatible with making massive equipment for infrastructure. But 3D printing allowed engineers and designers to produce production quality components within days or even hours of a design file being created. GE's innovation cycle still wasn't as quick as, say, Apple's near constant stream of updates to its operating system, but it is remarkably shorter than it had been. Traditional manufacturing at the industrial level needs months to produce testable prototypes.

As GE becomes a huge technology as well as an infrastructure company, its investment in 3D printing will help it in another unexpected way: making the company more attractive to younger employees. In front of an audience at Facebook headquarters, Immelt admits that manufacturing has lost some of its allure over the past few decades. "When I joined GE, 95 percent of all of you would think about going to work at GE. Today, maybe it's 60 percent. That's just life. So we have to be setting up shops to collect some of you guys when you might be thinking about making a change."

Trying to attract the talent that could choose to work at other tech companies was part of the impetus for building the San Ramon research center. But 3D printing doesn't just change where you work; it changes *how* you work. Instead of welding in the dank Rust Belt factories that, as Greg Morris remembers, "were hot in the summer and cold in the winter," additive takes place in what is essentially a laboratory. Medical implants or engine brackets are produced in a machine that quietly hums down the hallway. This type of manufacturing may be even quieter than a tech start-up.

Immelt also argues that new manufacturing techniques are making both GE and the United States much more competitive than they were a decade or two earlier. "By 2020, probably 25 percent of the parts that are done inside a high-tech GE are going to be done with 3-D printing, advanced manufacturing, and the like. . . . We can basically make every product GE makes and do it economically in the U.S. today."

Millennials have shown a strong preference for organizational cultures that are more like the horizontal, agile, "connect and inspire" model as compared to the "command and control" structure Boomers grew up with. In other words, they prefer processes that resemble manufacturing disrupted by 3D printing rather than traditional industrial techniques.

Within GE, the cultural impact of 3D printing was unexpected, but welcomed. "Our investment in 3D printing started because if we wanted to be responsive to customers, we knew this technology had to be in our portfolio," Furstoss says. "Then we realized that we could actually disrupt our culture through a manufacturing technology."

3D printing made FastWorks' goals realizable in a company where *The Lean Startup* processes were initially challenged by GE's fundamental organizational structure. "It's not the way GE is wired. It's not the way we're measured. It's not the way we think. But a bunch of us wouldn't let it go," said Furstoss. Then the two processes interlinked. 3D printing was impacting the company "at the same time as the GE FastWorks initiative. We wanted to be closer to our customer. It just all started to marry itself." Suddenly, acting more like a nimble tech company no longer seemed impossible, "because of the culture, collaboration, speed, new ecosystems—even more skilled workers in factories."

No matter how hard GE pushed culture change and FastWorks, it would not be able to fully transform itself into a nimble technology company without 3D printing. Even if employees realize that there is a different, better way to do things, without a major shake-up, the larger processes won't budge. By design, 3D printing messes up an organization's linear structure. In so doing, it disrupts the structures that are constraining the entrepreneurship of the company. 3D printing remakes manufacturing from its roots. This is what makes the technology unique from other forms of advanced manufacturing

"No one's ever thought of manufacturing this way," says Furstoss. "They may, rarely, think that the assembly line was a disruptor. I think in history industrial 3D printing will not just be remembered as a new manufacturing technology, but for its impact on company culture, and its creation of new opportunities and new business models."

Furstoss says that it would be easy for a company that makes physical products to ignore the disruption that's happening in web-based industries, but it would be a mistake. The value of additive goes well beyond making mind-boggling lattice structures.

The technology offers GE radically different possibilities, even for products that are currently built using traditional mass production. Instead of beginning a designing process with a list of things that you couldn't do, complex geometries make virtually anything possible. Instead of working with the same materials, additive has driven the creation of super alloys. Instead of simply designing a product and sending it downstream to the next department, additive has promoted collaboration at every step, releasing GE from the necessity of working within the same linear, hierarchical industrial processes that have dominated industry for three centuries.

Was that the intent of adopting additive-manufacturing technology? Obviously not. GE was primarily investing in a manufacturing process that would allow the company to be faster and more responsive to customers. Without any initial awareness of this transformation, though, the teams working with 3D technology shifted to an entrepreneurial culture. 3D printing made natural something the company had been trying to force others to do mostly against their will. It better aligned their work with the larger needs of a modern, innovative company.

NETWORKED DEVELOPMENT

In 1943, Columbia University professor Abraham Maslow proposed a new approach to the study of human development. Maslow criticized psychology of the day for being too focused on subjects that were neurotic or mentally ill. He proposed correcting this approach by studying what he called "exemplary people," such as Abraham Lincoln, Eleanor Roosevelt, and Frederick Douglass. His more positive approach was called humanistic because it emphasized positive human development.

Maslow's most lasting contribution was his description of what he called a "Hierarchy of Needs," which is often displayed as a pyramid. At the bottom are the most fundamental physiological needs, such as air, water, food, et cetera. But as the hierarchy goes up, human needs move from subsistence to higher-level development. The next level is safety, followed by love and belonging and self-esteem. Each stage has to be reached before a person can advance. For example, a human cannot feel love and belonging without having safety. A sense of safety is impossible without having enough to eat and drink. At the very top of the pyramid, just above self-esteem, is self-actualization. The goal of life, according to Maslow, was to strive for betterment, eventually reaching this top stage.

As is true with just about any mid-twentieth century academic work on human thought and values, Maslow's hierarchy has since come under criticism. But his theories on self-actualization also remain as foundations for sociology and management training. What is interesting is how his theories of human development line up with the advancement of technology from its First to Second Wave.

For example, the First Industrial Revolution dropped prices for manufacturing items by 90 percent. Suddenly huge numbers of people saw many of their basic needs met. They had more food, more clothing, more consumer goods. Trains allowed for free movement of labor, eventually increasing the standard of living dramatically. The printing press fueled the Enlightenment and Scientific Revolution, improving medical practices. Humans became free of scourges like polio and influenza pandemics. Cancer was more likely to be treatable. But these First Wave technologies also trapped many people in a mechanistic world. Success was measured in operational efficiency, creating jobs that were rote and repetitive. Most people in industrialized nations had enough to eat. They could feel love and belonging through providing for themselves and their families. But the same work that provided basic needs constrained them in their efforts for betterment. They were not self-actualized.

The automobile, the Internet, and, now, 3D printing are all Second Wave technologies with a profoundly different impact on human development. Think, for example, about the impact of the subway versus the automobile in Maslow's hometown of Brooklyn, New York. In the first half of the twentieth century, the subway expanded from Manhattan into Brooklyn and New York's other boroughs. Suddenly, the million people living in what had seemed to be a distant city across the river could travel rapidly to work and shop. Tens of thousands of people gained access to a whole new set of better-paying job opportunities. But the subway was also a hierarchical system with Manhattan on top. People in Brooklyn, Queens, and the Bronx often had to go through Manhattan to get anywhere else in the area. There is nothing particularly sinister in this organization. The New York subway map simply describes a lived, daily version of a command and control, hierarchical organization. If you lived in the outer boroughs on the bottom of the transit hierarchy, the subway didn't flat out deny you opportunities, but it made the best-paying employers relatively harder to reach. And you had a long, repetitive commute to get there.

Then along came the automobile, and a network developed around where people wanted to go. Highways and bridges sprung up, rapidly connecting the boroughs to each other. Whole new ecosystems and new jobs developed where it would have previously been impossible. In 1916, business opportunities outside of Manhattan were primarily local—shops, dry cleaners, and such.

The larger, more varied and profitable work was all centered in Manhattan.

A century later, the automobile has adjusted this balance. In a recent article in *Inc.* magazine, New York's ten fastest growing companies were asked to say what they loved about New York. The companies were located in all five boroughs except Staten Island. The Manhattan-based businesses were likely to talk about their location as a "premium address" or a brand that "communicates that we are innovative." Meanwhile, businesses in the Bronx and Queens immediately mentioned transportation access: "Our location gives us easy access via highway to everywhere in the region" and "You can't have mail trucks coming in and out of Manhattan; it would be a mess." Without the automobile, several of the fastest growing businesses in New York would not exist.

The point is that even a subway system with 722 miles of track isn't a perfect fit for every person or business. Automobiles opened up a whole new set of possibilities that allowed people in the boroughs to create new businesses. Just as Americans of an earlier generation took umbrage at having their possibilities defined by the often arbitrary acts of distant railroad executives, so New Yorkers found themselves chafing at the limitations imposed by being wedded to the nearest subway stop. Maslow's first-tier needs (transportation of any kind) yielded to second-and third-tier desires: choice, customization, a say-so in the daily decisions of life such as where to shop and eat—or pray.

Today, GE is walking its employees up the same ladder of human development. Until recently, the organization had been organized to maximize operational efficiency. Employee performance, as noted, was force ranked, with the bottom performers in every department often eliminated. The company was very successful within these metrics, but collaboration and innovation, both high-level needs, were generally de-emphasized, if not discouraged.

Then the organization began introducing 3D printing. Like the automobile, the technology encouraged customization. It could cost-effectively create low-volume variability. Suddenly, people were far more likely to be rewarded, professionally *and* personally, for experimenting. Designs that needed improvement were not failures. They were the natural result of low-risk experimentation. Likewise, the ability to work with digital files and create multiple-possible versions required collaboration. It brought the entire value chain into the conversation. Everyone from design to tooling to manufacturing to supply chain to operations to repair jumped in. Instead of a strict hierarchy in the organization, a collaborative network developed at GE.

Finally, with so many opportunities available to engage in the process, employees took more pride in their work. Instead of a cog in a machine, they were more likely to feel like they were part of a bigger effort. Machinists, for example, were being asked for their opinions. They gave input into questions they had never been asked before. Freed from the linear workflow, teams started working with a sense of purpose. They were reaching Maslow's higher-level development: esteem and self-actualization. But there was still one step to go.

BEYOND SELF

Later in life, Abraham Maslow issued retractions in exactly what self-actualization meant. He had begun to think that it was not merely the pursuit of individual excellence and self-betterment. Self-actualization needed an external dimension to it as well. People needed to feel part of a larger whole, in touch with those around them. At the highest level, this meant making a difference in the world. 3D printing has a role to play here as well because, as we've seen, its disruptive function extends well beyond manufacturing itself.

The natural world, for example, is a community to which we all belong. It was never entirely pristine, but the Industrial Revolution caused massive damage to its ecosystems in just a few centuries. Most environmental problems on earth—from polluted streams to dirty air to global warming—can be directly linked to mass production. This system of manufacturing has one goal: make as much as you can of the same thing as cheaply as you can. Everything else is peripheral. Traditional manufacturing's main byproduct is waste. What's worse, the heavily industrialized West accounts for just 15 percent of the world's population but 50 percent of consumption and waste. As the growing middle classes in China and India demand a more resource-heavy Western lifestyle, this will put an even heavier strain on global resources.

In the case of aerospace engineering, about 80 percent of the material bought to build planes doesn't ever take flight. But manufacturing with 3D printing's additive technology results in very little waste of raw materials. In some industries, 3D printing could reduce raw materials used by 50 percent. In many cases, the amount of water needed in production could be reduced by 90 percent. These would be huge benefits to the environment that has been so damaged by industrial production. The supply chains that distribute all these things around the globe are also incredibly wasteful. With 3D printing, only the raw materials get shipped long distances. 3D printing is not designed to be environmentally friendly. With current technologies, it still uses lots of electricity and raw materials. But the truth is that its primary processes are profoundly less wasteful. For some of us, this could mean manufacturing work that has a more positive connection to the natural world.

3D printing also encourages a completely different type of self-actualization: enabling us to have a direct impact in our own world. By its nature, mass production is bureaucratic and top-down. The structure was everything. A worker on Henry Ford's assembly line didn't make a car; he inserted bolt #49. Even people in upper management can't help but feel like cogs in a machine.

It wasn't always this way. Before industrial production, we were more likely to have a skill at creating something physical, be it a potato or a shoe or a cathedral or a plate. We solved problems and fixed things without going to the store. For most of human history, we were a world of creators and builders. Now we can see 3D printing lowering the entry barriers imposed by mass production. In the Second Wave of manufacturing, we will be able to break out of these constraints. We will be encouraged to create outside of fixed roles, or even on our own. 3D printing allows us to become a nation of innovators and builders again.

9

THREAT MULTIPLIER

The greatest fear in life is not of death, but unsolicited change.

—RAHEEL FAROOQ

3D PRINTING IS CHANGING THE world—and it will not always be pretty. The technology will cost some people their jobs, companies their profits, and nations their competitive edges. It will also have unforeseen repercussions, some of which could negatively impact fields as divergent as intellectual property, the environment, and bioethics. 3D printing is a powerful, transformational technology. As its use has become more commonplace, it has provoked debate, criticism, and even some fear. While many things about 3D printing are novel, fear and distrust of technology are not among them. Technophobia is as old as technology itself, but it is also a misunderstood phenomenon.

One of the most famous stories of technophobia begins in the textile manufacturing center of Nottingham, England. In November of 1811, fueled by a fear of technological progress, angry workers led by Ned Ludd broke into a factory at night and smashed the machinery. Over the following weeks, Ludd and his growing band of followers, aptly named the "Luddites," led other attacks throughout the heart of England's textile industry. Ludd's nighttime attacks and descriptions of his ghostly appearance gave the whole Luddite movement a sort of secretive ghoulishness that further terrified factory owners. The British government was obsessed with stopping the Luddites. Parliament made smashing machinery a capital crime. Agents were sent out with one goal: to capture Ludd. They all failed, for one very good reason. He didn't exist—and neither did his coordinated army of technology-hating warriors.

Ludd's name was in fact appropriated from the story of an apprentice who smashed his stocking frame after being reprimanded by his boss. The loosely banded Luddites adopted the name as a rallying cry, but they didn't hate technology. Their real demands were more familiar: better wages and proper training at their jobs. In fact, the nighttime attacks only started after soldiers broke up more conventional daytime labor protests. The workers didn't fear technology, but the change it represented in their stable world.

The same is true of 3D printing. People don't actually fear "the ability to print anything in layers." They fear the sweeping change it represents. But the promise and threat of new technologies always ride in on the same wave. As with the positive developments, we can't even begin to predict all the negatives. What we do know is that all disruptive technologies promote innovation. Here are a few areas of concern that are already being raised related to 3D printing, and some possible solutions to these problems.

ISSUE: INTELLECTUAL PROPERTY

In the 1980s and 1990s, record companies got fat off the sales of compact discs. The new format allowed them to transfer their entire back catalog to digital sound quality and then resell it at a big markup. But at the end of the 1990s, the booming business model for CDs was showing some cracks. File-sharing sites like Napster allowed people to easily upload and download the digital content on CDs. Instead of buying the music, fans could just rip them off the web for free. The Internet was the enabler of long-distance file sharing. And, unlike analog cassettes or records, the CD made music a convenient digital file. Combined, they made record companies' content easy to steal and share.

The rise of 3D scanning and printing poses a whole new IP risk. Everything in the physical realm can be reduced to a digital design file and then printed out anywhere in the world. Instead of ripping Tina Turner, we can rip a turbine blade. Does 3D printing make everything in the physical world just as vulnerable to IP theft as movies and music are now? Will 3D printing devalue intellectual property rights the same way that ripped CDs devalued musical content? Probably not, but there are plenty of lessons to be learned in how digitization of content affected record companies.

First of all, the record companies made plenty of mistakes in the face of this new threat. They talked about setting up their own iTunes-like platform to sell music online, but ended up fighting about what kind of file formats to use. Today, a consortium of companies in the 3D ecosystem is already working together to create a manufacturing file format that solves some IP problems. This format, called 3MF, would be harmonized between major players, including Autodesk, 3D Systems, Stratasys, and Microsoft. Unlike the current industry standard, STL, it would allow authors to inject IP rights information right into the 3D file.

Second, printing real end-use parts is still relatively hard compared to sending your friend an MP3. This means that 3D piracy will not be nearly as swift or devastating as it was for the record industry. Getting the CAD files to some object doesn't automatically allow you to reproduce the part in consumer-grade form.

That's not to say that 3D IP theft isn't a current danger for some companies. Gentle Giant is a company that sells limited edition action figures of everything from *The Walking Dead* to *Harry Potter* to *Star Wars*. In 2014, when it was acquired by printer manufacturer 3D Systems, Gentle Giant was looking at losses of at least $100 million in the next four years due to 3D piracy. The company's products, small plastic items, could be printed relatively easily and cheaply. But, unlike most small plastic objects, they also retailed for up to hundreds of dollars. In other words, they were in IP theft's sweet spot. So, yes, companies that make expensive branded plastic trinkets may be in trouble. But manufacturers of refrigerators probably don't have to worry as much. Stealing and printing these designs is not cost-effective yet.

Another mistake the record companies made was attempting to shut down file sharing through force. They tried legal action. This didn't work because a new sharing site could be set up just as cheaply. They tried introducing new super-high-fidelity CDs that couldn't be ripped and uploaded. That failed, too—consumers didn't want to buy them. Finally, a tech company provided a solution. Steve Jobs invited record executives out to Cupertino to show them his new application, iTunes. It was simpler, cooler, and better than anything else that existed. Many of the executives folded on the spot, turning over the digital rights to their massive back catalogs to iTunes with little negotiating power on the terms and royalty rates.

The record companies had tried to control and stop a digital leak. They ended up with low digital royalty rates. By contrast, Apple embraced the Second Wave, building an elegant interface that allowed people to easily and relatively cheaply download high-quality music files. Today the company generates $18 billion annually from iTunes. With its purchase of Gentle Giant, 3D Systems is pushing forward a similar business model for the branded action figures. Instead of manufacturing and shipping action figures, the company is making the designs easily available for printing on its consumer-grade printers. You can print the custom designed Han Solo in Storm Trooper disguise at home, but you need to download it from 3D Systems and print it on their Cubify printer.

The reality of the world today is that it's virtually impossible to stop people from 3D printing Dumbledore figurines. Once you accept this basic premise, you might as well simply try to get a piece of the market for digital downloads. The best way to cut down on piracy is not by trying to stop it, but by monetizing an easy and convenient way to print your products. In the process, companies will also create a new revenue stream.

At Fast Radius, for example, we are working with companies who want to create digital inventory that can be stored in the Cloud and used to generate spare parts on-demand. Today, companies are exploring digital inventory primarily for their own internal use, but this system could easily be opened up to consumers. This iTunes-like system would allow anyone to access a part they need, pay for a one-time print license, and then manufacture the part at home or at their local 3D print shop.

ISSUE: THE ENVIRONMENT

In the 1930s and 1940s, citizens of Los Angeles started noticing a strange yellowish haze floating around them. The thick, oppressive air caused teary eyes, a burning sensation in the throat, nausea, and headaches. Some days, the air was considered so unhealthy that kids were kept inside school during recess. Eventually it became so thick that drivers flicked on their headlights during the middle of the day and Los Angeles' airport was closed. The term "smog" was coined to describe this mixture of smoke and fog—but no one knew what was causing it. The city tried shutting down some factories and closed a tire production facility. They banned burning garbage. Still, the smog got worse.

Then, in November 1949, the California state agency charged with tracking the mysterious haze received hundreds of reports of extreme symptoms following a Washington State versus University of California football game. The area where the reports originated, Berkeley, was nowhere near any factories or other industrial polluters. In fact, the only machines present were the hundreds of cars sitting in bumper-to-bumper traffic. Scientists at the agency wondered: Could the smog be related to automobiles?

Of course, it had everything to do with the cars. The automobile was invented in an era when "environmental impact" wasn't a consideration to anyone. In 1949, cars had absolutely no environmental controls and were also incredibly inefficient, dumping huge amounts of unburned fuel from the engine directly into the air. A toxic mix of hydrocarbons, carbon monoxide, and nitrogen oxides shot out of the tailpipes. It wasn't until the 1970s that automakers were compelled to address the issues. Within a few years, they installed a relatively simple technology on their tailpipes: the catalytic converter. The exhaust coming out the back of autos became 99 percent cleaner. More recently, auto companies have had to reduce greenhouse gas emissions. They have invested in higher efficiency and electric vehicles. Again, the results are expected to be very quick. In a space of fifteen years, passenger vehicles in the United States will cut their greenhouse gas emissions in half. Like automobiles, 3D printing also provokes some environmental concerns, but technological innovations could begin moderating them quickly.

For one, 3D printing is very energy intensive. MIT has done research that shows 3D printing can currently use up to one hundred times or more the energy per part versus injection molding. This is a huge differential. If we suddenly made everything with 3D printing, we'd pollute the world and use up all of today's energy sources in the blink of an eye.

Of course, we only make a tiny percentage of parts using 3D technology. And where it is used today, it is cost-effective. This is because, while the per-unit energy cost may be high, producing the four units you need versus mass-producing a minimum of five hundred or one thousand results in lower energy usage and much less waste.

The extra energy usage in 3D printing is also primarily due to heating of the material. Today some of the biggest advances are coming in new materials, including materials that do not need to be heated for printing. In the future, we can safely assume that new material innovations will lower the energy usage associated with 3D printing dramatically.

Some of the extra energy usage created by the 3D printing

process is already offset by the massive efficiencies it creates. In aviation, for example, technical improvements are creating enormous savings in jet fuel consumption. But 3D printing's innovation could also result in energy breakthroughs. Today, the single biggest problem with bringing more solar and wind power online is efficiently and cost-effectively storing the energy generated. 3D printing is already being used to make batteries. If the technology leads to a breakthrough in energy storage problems, the result would end up being enormously beneficial to the world.

Another environmental concern is the large amounts of plastics used by 3D printing, some of which is non-recyclable. Once again, research in novel materials, as well as recyclable bases like metal, ceramics, and glass, can provide new solutions. Because of existing environmental demands, we will also likely move away from plastic print substances toward fully biodegradable building materials.

ISSUE: JOBS

3D Printing is just one of dozens of new advanced manufacturing technologies that threaten to replace or eliminate jobs, primarily for blue collar workers. There are few potential threats that generate as much heated debate. Fortunately, we have a growing set of evidence from other historical disruptions that can shed some light on this topic.

When automobiles displaced train service in the twentieth century, people also lost jobs. Work as, say, a rail engineer or porter was harder to find. But the expansion of the automobile created jobs everywhere, from salespeople to auto body technicians to traffic cops to owners of gas stations, to name only a few. Rail's efficiencies squeezed out labor costs. It took only one engineer and a limited number of other employees working in a constrained area to serve a huge number of people. The automobile's customization of transportation brought people and jobs everywhere. Until recently, information was also a pretty closed system. A small number of people created and produced, say, books and television for a huge number of consumers. Today, the Internet has created a whole new slew of jobs around the customization of information. McKinsey reports that another Second Wave technology, the Internet, has so far created 2.4 jobs for every one that was lost.

As 3D printing expands, certain types of manufacturing jobs will certainly be displaced. But much of this conventional production work is consolidated in giant manufacturing facilities where labor is increasingly being replaced by robotics. Customized production needs many more people to be involved in the process. And it needs people who are engaged at a more creative level than assembly-line production demands. Ultimately, 3D printing's most surprising economic impact will be as a net jobs creator.

ISSUE: BIOPRINTING

3D printing's rapid impact on medicine has reignited many ethical debates. For example, in 2013 researchers at Hangzhou Dianzi University in China printed a small working kidney. It only lasted four months, but it prompted conversations about the ethics of bioprinting. Today, the company Organovo is also printing liver and eye tissue cells.

It's easy to see the benefits of these technologies. There are long waiting lists for organ transplants—as well as real risks that bodies will reject the transplants. By printing the organs directly from a patient's stem cells, both of these problems can be eliminated.

So far, so good. It's hard to imagine an argument that people shouldn't be allowed to use their own stem cells to replace a failing organ.

But how about patients custom-ordering super organs? Scientists have already suggested mixing human stem cells with canine muscle cells to create enhanced organ tissue. What are the implications of mixing humans and dogs? What about muscle tissue that doesn't become fatigued as quickly? Or lungs that have a greater capacity to oxygenate blood? Should we be using medicine to create super humans? Should these people be allowed to compete in athletics? And how do you catch cheaters if the tests used to detect steroid abusers no longer work?

While these technologies are new, the outlines of the debates are not. Mary Shelley's *Frankenstein* raised many of the same issues nearly two centuries ago. Fortunately, printing tissue at home is still very hard. It's unlikely that these medical technologies will be used by off-the-grid mad scientists soon. More recently, ethicists have brought up concerns over cloning organs—or even entire human bodies as organ farms. None of these, or hundreds of other questions, have anything to do with 3D printing. The concern generated by 3D printing is not so much about the technology itself, as the effort to conceive of what change means.

What is certain is that 3D printing has had amazing positive impacts on medical treatment, from more successful knee replacements to more affordable prosthetics to life-saving organ transplants. It also allows drug companies to run tests on human tissue instead of clinical trials on humans or other animals. 3D printing provokes legitimate concerns. But they will have to be resolved in a way that maximizes the technology's beneficial uses—just as hundreds of other bioethical issues have been.

ISSUE: 3D PRINTED DRUGS

Lee Cronin, a chemist at the University of Glasgow, is developing a 3D printer that can mix, transfer, analyze, and purify medicine. His goal is to use his "chemputer" to democratize medicine. Patients could print their own medicine from a chemical file they got from the drugstore. But Cronin's goal of reducing drugs to basic building blocks has set off warning bells. Wouldn't the chemputer hypothetically allow DIY chemists to produce cocaine or deadly ricin powder? At some point, possibly, but possessing both would still be illegal and tightly regulated. Throughout history, governments have been surprisingly successful in keeping up with new technologies in the interest of consumer protection and the common good.

On the other hand, Cronin argues that his invention would have a massive positive impact in the poor nations of the world. Countries in Africa, for example, often have relatively poor access to pharmacies. The relatively cheap chemputer can print any pill locally. They could be set up as micro-pharmacies across a country to deliver life-saving medicines, saving and improving tens of thousands of lives while raising living standards and productivity.

3D printing food may also have a similar impact on malnutrition. Unfathomable amounts of food spoil every year, much of it designated for developing nations. Almost all of the spoilage is due to inefficient supply chains. This may change due to the food version of the chemputer. 3D printers can divide basic food elements into printable materials. These elements can be placed in containers that preserve them for years. The machines can print enjoyable and nutritious food in areas where it is needed most, offering taste and texture variety far beyond what is currently available. The end result of locally produced drugs and food would be much healthier populations. For the developing world, 3D printing could quite possibly offer a break from some of the worst impacts of poverty.

ISSUE: NATIONAL ECONOMIC SECURITY

Around the world, the Great Disruption caused by 3D printing will shake up not just individual industries but entire national economies. For example, China is vulnerable to some of 3D printing's future impacts. Its huge economy is largely based on mass-producing parts at low per-unit prices. But the slow, inflexible, and expensive supply chain that connects it to import nations adds costs right back in. By offering alternative solutions to supply chain issues, 3D printing cuts down the total cost advantage of offshoring production. 3D printing's capacity for localized production is also likely to reshore some manufacturing jobs in the United States and other wealthier Western nations. Is this a disaster in the making for China? Will this create losses that ripple out into the global economy?

Not necessarily. Chinese leaders already realize that their current export model is not permanently sustainable. Manufacturing jobs are beginning to migrate to even lower-wage nations like Vietnam. China has already invested huge resources into creating their own 3D printing industry. Rather than hurting China, the technology will likely allow it to naturally pivot from an export-based economy into an economy based on domestic consumption over time.

Likewise, wealthy countries like the United States and EU member states will benefit as some 3D printing will make reshoring increasingly attractive. At least some lost manufacturing jobs will return home. But 3D printing also poses a threat to these same nations. Today, countries like Germany, Japan, and the United States enjoy a huge innovation advantage over most of the world. Like it or not, 3D printing will level the field.

Remember Arie Kurniawan, the Indonesian engineer who won GE's design competition? There are thousands of Aries out there, most of them in Asia and other parts of the developing world. Over the next few decades, tens of millions of new innovators will gain access to customizable and complex prototyping devices and even production capabilities. The distance between third world entrepreneurs and Ph.D.s working for wealthy multinationals will decrease. 3D printing encourages innovation across the board. If they lose their focus, companies in the developed world like BMW, Nintendo, and Apple will find it very hard to maintain their dominant positions in product design. GE's Chief Economist Marco Annunziata says that the wealthiest nations will be benefactors of 3D printing's global impact—except those that fail to innovate. An unwillingness to experiment with the new technology will lead to backsliding versus the rest of the world.

As with other concerns, 3D printing will be a threat and an opportunity within the global economic order. But the technology will primarily act as a global leveler. As it reshuffles capabilities, 3D printing will provide people in the advanced and developing worlds alike with previously unattainable opportunities.

AFRICA AND THE *NEW* WEALTH OF NATIONS

For more than a century, much of the continent of Africa has been locked in an impoverished grind. During that time, the world's global flow of goods and services has been governed by a commonly accepted orthodoxy: All countries benefit by producing whatever they can most efficiently, while trading with external countries for other goods or services. As we have watched this theory play out, the wealthiest nations have benefitted by investing heavily in manufacturing infrastructure, while the poorer countries that exported raw materials or agricultural goods have gained little. In fact, sticking to what they can produce most efficiently has left African nations overwhelmingly locked into poverty. The truth is that, short of stumbling upon massive oil reserves, countries without manufacturing sectors have remained very poor.

This industrial order is felt devastatingly across Africa, where most countries have little to no production capacity. Until now, they have had almost no hope of ever developing an industrial base. They essentially face the same entry barriers as a manufacturing entrepreneur. Mass production requires enormous start-up capital and infrastructural investment. In the future, these nations will benefit in the same way 3D printing enables entrepreneurs. In a sense, developing a manufacturing sector in an African country is the ultimate start-up, but it's no longer the ultimate pipe dream.

For example, traditional manufacturing requires that multiple parts, including prototypes and molds, be shipped on African nations' dicey infrastructure.

Because 3D printing is more local and self-contained, it allows African entrepreneurs to overcome some of these problems. An appliance-sized 3D metal printer is relatively small, compared to mass-production machinery. A printer just needs raw material and electricity to manufacture. With these simple needs met, it can print out the same extremely complex design just as easily in Ghana as in Germany.

3D printing's disruption of African societies may look a lot like cell phones' impact over the past several decades. Just as African countries have insufficient roads and ports, their communications infrastructure is also lousy. The vast majority of people have never had any home phone access at all. This is not just because of poverty, but because wires often don't travel outside of cities. Or, where they do exist, the lines may wait months for repairs. Then cell phones arrived on the continent. Mobiles still need an infrastructure, but relatively speaking, their demands are low. You don't need to be literally wired to it. You just need to be in the right vicinity. Like 3D printing, they are a more flexible technology.

The result was remarkable. Tens of millions of Africans who never had landlines—or any kind of electronic communication—got cell phones. Much of the continent will never be wired, but now it doesn't matter. Many Americans are getting rid of their landlines as well. With cell phones, Africans successfully bypassed a whole technological era. They moved directly to wireless communication.

3D printing provides African countries a similar leapfrog opportunity. The technology allows an entire continent to jump past centuries of industrial manufacturing and straight to 3D printing. They would go from having no heavy industry immediately to the complexity and customization of digital manufacturing.

Nigeria is a case in point. Africa's largest economy and the eighth largest country in the world by population, Nigeria has never developed a significant manufacturing sector. This failure is due to endemic corruption but, more importantly, to the lack of investment in roads and other infrastructure necessary to bring parts in and out of the country.

Early exposure to 3D printing in Nigeria has produced promising results. In 2014, GE Garages—traveling showcases for the company's latest manufacturing and prototyping technology—decided to take its exhibition abroad. The first stop was Lagos, the economic capital of the country. At the event, staff technicians and several young local entrepreneurs collaborated on inventions.

One Nigerian inventor, Saheed Adepoju, used 3D printers to develop his prototype, the country's first tablet computer. A teenage entrepreneur also used the technology to develop an invention that used urine to both power a light and create clean water. Drinking water and electricity access are both in demand across the continent. A cheap, reliable source of both would be transformative for neighborhoods, cities, and even countries. Health and productivity could be improved—and people would have a nearby place to charge their cell phones.

Nigerian manufacturing innovators may even have an advantage in readily adopting the new technology. Since they never learned traditional mass-production engineering, they don't have to unlearn its design constraints. Today, some engineering students in Nigeria and other parts of Africa graduate knowing how to design only for 3D technology. This could help them leapfrog the industrial era and make the country an innovative manufacturing center.

GE's Marco Annunziata is one of those who think that developing nations like those in Africa will be "primary beneficiaries of additive manufacturing." One natural advantage Nigeria possesses is resources. The country is home to a treasure trove of valuable raw materials—oil, coal, tin, columbite, and rubber. The same is true of much of the rest of Africa as well as impoverished South American countries like Bolivia. They are rich in extractable minerals, but they can't produce finished products. They can't turn their enormous natural wealth into finished products. Instead, they sell off mineral rights to global mining companies, gaining a fraction of their potential income. With 3D printing, local businesses could actually produce value-added products for export.

THE UPSIDE OF THE DARKSIDE

3D printing has a dark side. The very same reasons that make it exciting also make it dangerous. 3D printing promotes innovation by putting the technology in people's hands. But it also creates IP problems by blurring the line between producer and user. It enables manufacturing entrepreneurs to get their start-ups off the ground. At the same time it makes it easier for terrorists to manufacture weaponized drones. Wherever there is such sweeping change, opportunities and threats will emerge. Opportunities need to be seized, and challenges exist for entrepreneurs to solve. Ultimately, 3D printing's biggest threat is to the status quo, and to leaders who are unable or unwilling to adapt.

10

DISRUPTIVE LEADERSHIP

The world is moving so fast these days that the man who says it can't be done is generally interrupted by someone doing it.

—ELBERT HUBBARD

IN THE EARLY MORNING OF March 19, 2003, two U.S. Air Force F-117 Nighthawks dropped four bombs on a compound where Saddam Hussein and his sons were believed to be. At that exact same moment, forty Tomahawk missiles were fired from at least four U.S. Navy ships located in the Persian Gulf and Red Sea—a cruiser, a destroyer, as well as two submarines. The following morning, CIA special operations commandos seamlessly infiltrated Iraq and called in additional targeted air strikes all around the country. U.S. ground troops followed, pouring in from Kuwait.

The U.S.-led coalition forces proceeded rapidly with their invasion plan. They secured oil fields in the south. They avoided major cities, and the Iraqi forces stationed there, in pursuit of their main objective: the Iraqi capital of Baghdad. Exactly three weeks after the invasion began, coalition forces took Baghdad and Iraqi President Saddam Hussein went into hiding. The invasion of Iraq was one of the swiftest and most decisive victories in all of military history.

On May 1, 2003, President George W. Bush landed on the deck of the USS *Abraham Lincoln*. The resulting iconic image has the president congratulating the troops while standing in front of a banner stating "Mission Accomplished." But the ringing sounds of victory would quickly fade. The war was far from over.

The subsequent U.S. occupation of Iraq lasted until December 18, 2011—or 150 times longer than it took to topple the government. During the invasion phase, 139 U.S. combatants were killed. During the ensuing occupation, another 4,253 U.S. service members lost their lives. These numbers mean that in the nine years following their overwhelming defeat of the Iraqi army, the U.S. military's occupying forces continued to die at a very high rate. When the soldiers finally pulled out, there were no presidential visits to aircraft carriers, no cheerful declarations of victory, no banners. And the Iraq they left behind was, after nearly nine years of nation building, unsettled at best.

How could this happen? It wasn't because, after the successful invasion, the U.S. military suddenly decided to take it easy. The same generals who commanded so brilliantly during the invasion didn't all of a sudden become less competent. The Iraqi forces didn't miraculously become formidable adversaries. How can a military force go from being such an overwhelmingly dominant competitor to being hobbled by irregular forces in a matter of less than a year? The answer has something in common with the decline of the department store.

IN 1886, RICHARD Sears was presented with a rare opportunity. While on the job at a Minnesota railway station, Sears noticed that a regional jewelry company decided to return an order of gold watches. Sears purchased the watches himself and sold them in smaller batches to other agents up and down the railway. The watches quickly sold out, making Sears a handsome profit. He ordered more watches, and soon the company that would later be named Sears, Roebuck & Co. was born.

The Sears company quickly became a breakout success, and Richard Sears was the guiding genius of the business. He understood his customers, predominantly farmers, and was a master at selling, advertising, and merchandising. By the 1920s the company was flourishing. In 1925 Sears opened its first retail store. By 1930 there were over 400. At its peak rate of expansion, Sears was opening an average of one store every single day of the year.

The company continued to grow its empire, undeterred by the Great Depression or WWII. In 1972, the Sears headquarters building in Chicago was the tallest in the world, three out of every four Americans visited a Sears location, half of U.S. households held a Sears credit card, and Sears sales accounted for 1 percent of Gross National Product. The retail giant seemed invincible.

A hundred and ten years after the Sears empire began, a young Frenchman living in the United States had a much different opportunity. Pierre Omidyar's girlfriend had recently returned home from Europe with a bunch of Pez dispensers. She wanted to sell some of them, but was having a hard time connecting with buyers. Omidyar thought that maybe creating a web auction would be a good "market mechanism" for setting the right value for the items.

He wrote some software to create an online auction site and went to the local government office to register the domain name echobay.com. It was taken. Reluctantly, he decided to abbreviate it. "What about eBay? Turned out to be my lucky choice," reflected Omidyar. The company took off almost immediately. For the first few years, eBay's computer servers had trouble keeping up with demand, crashing almost once per day. Soon, eBay's sales surpassed Sears, which was mired in a downward spiral. Today, Sears teeters on bankruptcy, while eBay has achieved a market capitalization of more than $27 billion.

How does a company like Sears that for decades wielded such formidable competitive dominance experience such a swift slide into irrelevance? Some say the department store's demise followed the decline of the American middle class. But this cannot explain the free fall of a retail giant while new companies like eBay aggressively grew sales. Sears had over one hundred years of sales, marketing, and merchandizing expertise. Pierre Omidyar had none. What were the changes in the competitive landscape that allowed such a great company to plummet to bankruptcy so quickly while others prospered?

FROM STATIC TO DYNAMIC

The Great Disruption is a book about the fascinating and daunting changes that 3D printing technology will usher in. To be clear: 3D changes nearly everything. But this book is also the revelation of an idea, and that idea is that the adoption of every new technology spreads in two waves, not one. Just as there are very distinct markers for each phase—mass production in the First Wave and the inevitable customization that follows in the Second Wave—so, too, do these phases represent very different competitive environments, and require significantly different approaches to leadership. As 3D printing technology ushers in the Second Wave of production technology, it is critical that all leaders take note, and adjust rapidly to an entirely new competitive environment.

The stalled military occupation in Iraq and the sudden demise of Sears both resulted from this dramatic shift from the First Wave to the Second Wave. Fighting insurgents in a desert doesn't seem to have too much in common with pushing perfume in a mall. But both stories share the same underlying pattern. First of all, these are clear examples of competitive landscapes that shifted from static to dynamic environments. No one who gauged things by twentieth-century-operational standards could have predicted that the U.S. military and Sears would begin to flail so quickly. This is because, in the First Wave, things don't change much over time. The competitors are clear. The rules of engagement are clear. Deliberate planning in the war room or the boardroom likely results in predictable outcomes in the marketplace. But in the Second Wave, this all changes. The environment becomes dynamic, with situations changing more rapidly than you can process them, let alone plan for them.

This explains why the Second Wave of the Iraq war, the occupation, was a different beast from the First Wave, the invasion. The invasion was a well-thought-out, well-planned, and nearly flawlessly executed strategy. The military knew where the enemy would be, could estimate the resistance, and could move its pieces around like a game of Risk. Similarly, for decades, Sears competed in a First Wave environment. It used its sales, marketing, and merchandizing expertise to decide for the consumers what they were going to buy. It studied the slow entry of competitors and reacted accordingly. In both of these First Wave environments, competition was static and the scales were greatly tipped toward the incumbent.

Not so in the Second Wave. Think about trains versus automobile-based systems. With rail, you can predict how many people will be on what train, and you know exactly when the trains will arrive. An auto-based system is made up of millions of people making millions of individual decisions in real time.

The U.S. military and Sears began to struggle after the occupation of Iraq and Internet retailing each fostered a Second Wave competitive environment. Instead of command and control military leadership, occupations are won and lost in dynamic competition. Once they began combatting networked, mobile, local resistance, the U.S. military needed to be using real-time information, not static assumptions, for competitive advantage. Similarly, Sears was trying to compete against a networked, nimble web platform with its traditional one-thousand-page marketing plans. eBay had virtually no plans at all. Instead, it had basic rules and tools that both sellers and consumers could use how they chose. The market set the price, determined who merchandized the best, and where and when the products would be sold. Omidyar and eBay didn't need to make these decisions; they simply had to watch and react to what was happening as quickly as possible. The Second Wave is about real-time decisions where much is out of your control and the value of planning is replaced by agility.

As we enter a 3D enabled Second Wave production environment, manufacturing will begin to operate in the same way. Linear production and supply chains will be replaced by dynamic networks with interacting nodes all around the world. Finite groups of powerful and predictable producers will be augmented with thousands or even millions of smaller new competitors. Stability will be replaced by disorder. As we shift from control to networks—the first distinguishing characteristic of the Second Wave—predictability will be replaced by chaos.

FROM ASSETS TO INFORMATION

The second distinguishing characteristic is that, in both cases, the physical assets that allowed the U.S. military and Sears to dominate the First Wave quickly lost value in favor of information. The U.S. military claimed victory in Iraq in much the same way that it won WWII. In 1940, President Roosevelt announced such an ambitious goal in increased production of military products—including 185,000 airplanes—that Adolf Hitler's advisers told him it was American propaganda. They had reason for skepticism: In 1938, the United States had made a total of 3,000 planes. But by the end of the war, American factories had surpassed Roosevelt's challenge, producing 300,000 planes. The United States' armed forces were initially sparsely equipped, but its factories made planes and guns and jeeps faster and cheaper than anyone else. The country eventually produced two thirds of all Allied military equipment used. This massive production of tanks and other armaments was a decisive factor in the most industrialized war at the time. William Knudsen, appointed lieutenant general by President Roosevelt to oversee war production, said "We won because we smothered the enemy in an avalanche of production, the like of which he had never seen, nor dreamed possible." Six decades later, the U.S.-led coalition dominated the Iraqi invasion based on assets again.

Sears' century-long market dominance was also based on assets. The company purchased prime retail real estate. It used massive volume to purchase retail items at huge discounts not afforded to smaller competitors. It outspent other retailers in advertising sometimes by as much as a hundred to one. In short, it used its massive assets to create overwhelming barriers to entry for future competitors.

In the Second Wave, this massive infrastructure quickly went from asset to liability. eBay didn't care about real estate. It operated only on the Internet. It didn't need expensive addresses, or tens of thousands of employees. Its growth was fueled by word of mouth, not by $2 million Superbowl ads. Sears could compete on expensive assets only if its competitors were paying the same price. eBay didn't. All of the sudden, Sears' expensive business model became a glaring disadvantage.

Similarly, having the largest arsenal of weapons holds a critical advantage only when fighting a traditional war. When the battleground becomes the unfamiliar backstreets of ancient cities, calling in artillery strikes is of no use. Iraqi insurgents were blowing up $250,000 Humvees with $50 homemade explosives. "Expensive" and "effective" took opposite paths.

First Wave manufacturing and supply chain competitors have long operated under the protective umbrella of capital assets. Entry into the marketplace requires building massive factories, long, complicated supply chains, and producing items by the millions. What chance do newcomers have to ever get off the ground? But in the coming Second Wave of production, these massive assets are largely negated. The elastic manufacturing cloud allows you to tap into the production levels you need, large or small. Production becomes localized, so sophisticated supply chains are no longer a price of entry. In the Second Wave, competition is based on who can access and use information, not assets, most effectively.

FROM HIERARCHY TO THE NETWORK'S EDGE

Finally, the strong, hierarchical structures that enabled these large organizations to move masses of people in lockstep in the First Wave contributed to their downfall during the Second Wave. The United States dominated the Iraqis through effective management. A Pentagon report found that the number one cause of the rapid and complete Iraqi defeat was Hussein's incompetent and inconsistent micromanaging of military plans. By contrast, the United States had maintained a clean and stable command and control structure. The American invasion was well planned and accurately carried out by the soldiers on the ground.

The advantage in command and control hierarchical management did not help the United States much during the subsequent Second Wave occupation. The U.S. military strove for logistical efficiency, moving massive amounts of people and things around the country and the world. But in Iraq, local insurgents operating in small groups or as solo operators repeatedly attacked American convoys on long desert roads. Speed became a critical asset. Rather than helping, the multilayered hierarchical structure of the military only slowed down the flow of information. By the time information had been passed to the top, and resulting decisions flowed back down, the competitive landscape had already changed dramatically.

Sears had built its own multilayered bureaucracy over time. When styles and customers and competitors changed much more rapidly, this large hierarchy simply created friction in the flow of information. Within large corporate organizations, hierarchies have another issue. No one likes to deliver bad news to their boss. Like the children's game of telephone where kids whisper a message that gets totally obscured by the time it reaches the last person, the reality of bad things happening right on the front-lines with customers rarely makes it up to senior leaders intact. Not only are they getting information from the front lines slowly, the information they get is not accurate.

Second Wave leadership requires information to come from the edge. This is a world where things trending on Twitter directly impact tomorrow's production runs—where products are customized in real time based on the unpredictable wants of individual consumers. As we will soon see, Second Wave organizations push decisions to the edge.

FIRST WAVE LEADERSHIP

In 1960, a buttoned-up marketing professor named E. Jerome McCarthy introduced the four elements of what he called a "marketing mix." Also known as the "4Ps," the categories provided a baseline for marketing for years afterward. The first "P," *product*, is something that customers are willing to purchase to satisfy some need or want. *Pricing* determines how the product is situated in the marketplace. *Place* refers to the environment or manner in which the product will be bought—online, big box store, niche boutique, airports, etc. Finally, *promotion* covers the elements most typically associated with marketing, such as advertising, public relations, and brand management.

———

THE 4PS ARE something else as well: a synthesis of First Wave leadership for much of the last century. You can imagine the executives smoking in a cherry-paneled conference room while discussing the four Ps. They interpret what the masses want, and then decide what they will get. They design the product, set the price, pick distribution, and plan out promotion. This leadership is absolutely not about networking. The decisions are either made with little feedback from outside the room or, if feedback is requested from a focus group, it is funneled into a hierarchical decision making process. There is little if any conversation with users, much less cocreation.

It's easy to pick away at the 4Ps, but ultimately they define what we often mean when we say "manufacturing leadership." The 4Ps are about making a bunch of stuff and then convincing people that they want it—the exact model that manufacturing still clings to today. Follow the money: Last year, the top twenty countries in the world spent $300 billion on research and development to design products for people to use. That's a lot of money. But it is dwarfed by the trillion dollars they spent trying to convince people that they wanted to buy all these products. Even back in the twentieth century, this sort of planning sometimes created epic flops. Think of Ford trying to get rid of tens of thousands of Edsels. As we transition to a Second Wave world, this paradigm will be even more likely to lead to failures. See, for example, how the U.S. military was saddled with the expensive job of convincing Iraqis that they wanted the American occupation.

The 4Ps work on the assumption of predictable market outcomes. Technologist Mickey McManus describes the phenomenon as similar to holding a pool cue from the tip, facing down. The First Wave is about trying to get it to stop swinging entirely. To make it static, to control it so there is no variance, it doesn't move, it's predictable. The 4Ps are a First Wave phenomenon, operating with static competition. It is about planning and hierarchical stability. It is generals moving tank figurines around a large map. When you have a settled field, you can focus on efficiency and eliminating variance to maximize predictability. It is Jack Welch's Six Sigma, relentlessly driving out variance in favor of predictability and control. In the First Wave, the goal is to reach equilibrium. This approach has worked successfully for decades, but today, it is the absolute perfect way to fail, from marketing to nation building. Instead you now need to understand how networks work within your company.

ON THE EDGE

A few years ago, I met with Steven Rice, the senior executive in charge of talent at Juniper Networks. In the mid-1990s, Juniper started as a provider of networking products. The company enjoyed success with its high-end Internet routers, stealing customers from industry behemoth Cisco Systems. In 1999, Juniper's IPO set a one-day record for trading in the technology sector, increasing by 191 percent. Since then, the company has continued to expand its sales both domestically and internationally. In short, Juniper must have talented managers who understand the company very well.

In his office Rice eagerly told me a completely different story. His team had just completed an internal analysis. The project had started as a relatively pedestrian effort to understand how they were serving one of their most important global customers. The team identified all the people involved and the different points of contact. The results were anything but expected.

"Look at this!" Rice leaned across his desk and handed me a chart.

I studied a series of dozens of lines crisscrossing a slightly amorphous space. At points they arranged themselves in fan-like constellations. The chart looked a bit like a map of an airline's global flight service.

"What is it?" I asked.

"Here, right here!" he pointed to a fan-like shape on the lower left hand side of the diagram. "You see? This person is a hub. She sits right in the middle of dozens of the most critical customer interactions."

Rice looked up, "But guess what—she's low in the organization. She gets relatively little oversight, compensation, or training. Her organizational value—from a distance—might seem really low. But her value isn't. She is a huge part of one of our most critical delivery networks."

Rice gestured again at the chart. "Okay. Another thing. Can you identify the customer executive on this chart?"

I glanced back at the diagram.

"Any guesses? I could tell you—but that's not the point. The point is that there is *no* central point of control on this chart. There is no discernible hierarchy."

"Now look at this!" He handed me another chart. It was a neatly laid out structure that showed authority flowing predictably from the customer vice president down through directors, managers, and so on.

"This is our organizational chart. It's the structure *we think* we are managing one of our most important customers with. But it bears little resemblance to how we are *actually* working with this customer. This has a hierarchy, a linear decision making process,

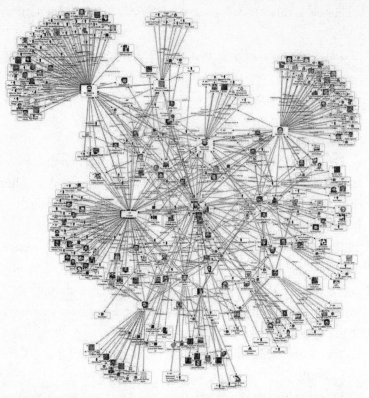

Example of a social network analysis graph. FMS ADVANCED SYSTEMS GROUP/
SENTINEL VISUALIZER

it goes from the top down and from the inside out. That's not what's happening at all."

What Rice's internal audit revealed was how the Internet, the Second Wave of the information age, has changed business in ways that were not envisioned when Juniper Networks started up. At the time, the Internet was a tool for sending email, doing research, and buying books or music. Then, over the next couple decades, the speed and amount of information sharing increased exponentially. To keep up with this rush of data, employees realized that they couldn't wait for answers to come back down the chain of command. So they took advantage of their enormous new access to tools and information to make fast decisions.

This poor fit between the management and employees sounds like a disaster in the making. But Juniper's audit had shown the opposite. The employees who were operating at least semiautonomously were very effectively managing one of the organization's top clients. Juniper was managing this customer virtually through a series of interconnected networks of people.

In fact, without its employees disrupting the organization's command and control structure, Juniper would likely not have performed nearly as well. Not only would employees have been waiting for responses further up the ladder, but the logic used by management to make decisions was outdated. As information and speed accelerate, we can no longer indulge a linear approach to problem solving. We've been trained to break down problems into manageable pieces. Today, organizations need to tackle bigger issues in real time. This in turn means that a chart of the traditional operational pecking order is becoming much less relevant. At Juniper, low-level employees are the ones making decisions now.

"Those at the edge of the organization are empowered to make the decisions, better and more quickly than in the past," said Rice. "They are so close to the customer that innovations are appearing without even looking for them. And the engagement level of these employees is as high as for any group."

We typically think of these command and control structures as up and down. In this case, it seemed like the junior employees toward the bottom were taking initiative. But Rice described it slightly differently. The chart he'd handed me showed that Juniper's decision making had moved "to the edge." In his rendering of the structure, command is centralized deep inside an organization. This is more like a galaxy than a flowchart. But describing employees as on the edge explains why they were so successful. These people were the most likely to be dealing directly with customers. They were the most in contact with the world outside the organization. They received a lot of information before anybody else. If they were looking for other information, they would be more likely to get it from another employee on the edge of the business.

Critical decisions made at the periphery, closest to and often involving customers; information gathered and shared at the edge and pulled from the core only when needed—this is a complete reversal of decision flows, from inward out to outward in. Organizations move from a handful of leaders at the core to massive collaboration at the edge. Competitive advantage is won or lost on the periphery. Success isn't based on ideas from the core, but originates from facilitating engagement at the edge. So how do you encourage this?

ITUNES VS. APP STORE

One example is the difference between iTunes and the Apple App Store. At first glance, both have a lot in common. Apple created two platforms. One sold music online; the other sold apps. In both, Apple made the rules and took a 30 percent cut of all profits. But the App Store is actually much more valuable for Apple. With iTunes, Apple makes money by selling popular content online. The idea and platform was innovative, but ultimately Apple is just serving as a web retailer. The company wants to sell lots of music, but it doesn't care if they make their money selling multiple genres—or exclusively chamber music and jazz fusion. As long as they can sell what's popular, the variety and quality of content doesn't matter.

By contrast, Apple benefits greatly by having a diversity of high-quality apps available through its App Store. If developers just made one kind of app, like games, the iPhone would be a fancy communications device and a mobile arcade. But the enormous variety of apps available has turned the device into a personal trainer, a fax machine, an encyclopedia of music, a photo studio, a navigator, a meteorologist, and a doorman. While diversity of music on iTunes does nothing extra for Apple or the iPod, diversity of apps radically improves the functionality of the iPhone. Would in-house Apple developers have created Storyful, an app that sorts through media to identify credible items? What about Photomath, which lets you take a photo of a math problem and then solves it? Maybe or maybe not, but they didn't need to. Instead the App Store relied on a huge range of innovators on the edge of a network to see the problems to solve, create the apps, and sell them. As a result, Apple has a huge incentive to develop a diverse network of developers.

With iTunes, Apple became the world's biggest seller of music, but its model is relatively static. It is essentially a storefront competing with other online sellers. For years, Amazon's MP3 service has been eating into iTunes' dominance. Meanwhile, streaming music services have eaten into online sales. With the App Store, Apple is constantly expanding not just the number of apps but the functionality of iPhones. This dynamic and varied ecosystem is a large part of why iPhones continue to increase Apple's market share among smartphone users.

SECOND WAVE LEADERSHIP

Second Wave leadership is like taking the same pool cue but, instead of holding it until it stops swinging, trying to balance it the palm of your hand. The environment is dynamic and constantly changing. It's still possible to make good decisions, but not by using First Wave management skills. Success requires constant information. Leaders need to make quick decisions. They must be able to change course. The goal is to interpret all these changes, all this information as quickly as possible, and make decisions to keep everything in balance. It's about agility, not control.

At Juniper, Rice was surprised to find out how information was actually flowing through his organization. This wasn't something Jim Collins' *Good to Great* would prepare him for. But he could recognize a good thing when he saw it. He didn't try to rein everyone back into an outdated First Wave leadership model. Instead he studied how it was possible that people at the edge of a network were able to serve Juniper's clients so well. In many cases, they were clearly making good decisions quickly. He celebrated their success and learned from it. The same thing needs to happen as the Second Wave, 3D production and mass customization, revolutionizes manufacturing and all impacted industries. The key: capturing information and using it faster and more effectively than your competition. But remember, there are multiple types of information at play.

CUSTOMER INFORMATION What is trending locally? What opportunities are there for customization and how do you collect this information?

PRODUCT INFORMATION What data can go into the initial design to come up with the optimal design? How can you create objects that collect data and communicate to each other?

PRODUCTION INFORMATION How many do you need to create or print? Where is the best location? Who has capacity now? This is like Uber—it is about knowing not only who is capable of delivering the service, but who has interest/capacity now. It is about knowing the right person at the exact right point in time.

VALUE INFORMATION What is the total cost of your production decision? Pricing should not be static, but based on the value produced. Knowing all the factors of cost and value are critical inputs driving the optimal production decision.

LEARNING ORGANIZATIONS

At the same SXSW festival in Austin where Oreo vending machines were popping out customized cookies, several other companies with Second Wave business models were very busy. The festival brings 150,000 visitors into a city with 30,000 hotel rooms. One result of this influx is a huge demand for accommodation funneled through services like Airbnb and HomeAway. Many residents hate the week-and-a-half occupation driven by the festival, but homeowners can at least turn a nice profit. Two bedroom apartments that might usually rent for $250 a night can often command double or triple that amount during SXSW, with some going for as high as $1,500.

Local car owners can also profit. These massive crowds drive surge pricing on ride sharing apps like Uber and Lyft. In some crowded areas at night, riders are paying five times the normal fee for a ride. You could call it price gouging, but these share-economy businesses have helped Austin accommodate crowds that would overwhelm the traditional local tourist infrastructure.

These companies are consummate Second Wave service providers. None of them has centralized production. They have a central platform, which sets the rules and recruits participants to their networks. But the decisions about supply and demand decisions are made entirely at the edge in real time. Pricing is driven by who is available, interested, and closest to you to provide the service. Neither Uber nor Lyft are involved in any such decisions. They simply provide information to enable customers and drivers to make the decision directly.

SECOND WAVE PRODUCTION

These lessons gained from web-based platforms contain valuable insights into what Second Wave production is going to look like—and how to lead it. Uber is a distributed transportation solution. Distributed manufacturing works the same way. The Oreo vending machine is a great example. It provided local, customized, even personalized, production. It pushed decision making on the final product from a boardroom directly into the hands of a consumer. To move the project forward, the team at Oreo had to fight against internal resistance at Mondelez. But after its massive success, it is being heralded by the parent company as the new way to drive innovation for the world's most popular cookie.

Likewise, Nike has begun imagining its shoes not as products to be purchased, but species to be grown. What if you could take LeBron XIII basketball shoes and customize them to the point that they were virtually unrecognizable as the base model? Nike wants to offer customers the chance to create shoes that are uniquely valuable to them.

SO WHAT SHOULD LEADERS DO, RIGHT NOW?!?

Let's look next in close detail at what you as a leader should be doing right now. As we have pointed to previously in the book, the two biggest disruptions caused by 3D printing are going to be centered on supply chain compression, and product and production innovation. To end up on the plus side of this disruption, you need to know your organization's strategic interests. The chart below is meant to help you do that. Locating your position relative to the two primary disruptive elements will help you gauge your risks and vulnerabilities in a 3D printed world.

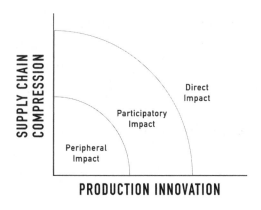

Take a look at the bottom of the y axis, which measures the impact of supply chain compression. This is where you'd find, say, financial services and Internet companies. These are companies with lower or virtually no volume of production and thus little exposure to direct impacts, at least in the short term. A little farther up in the participatory zone we find companies like GE and 3M. Their business involves some manufacturing, and supply chain is important—but it's not their entire business. Finally, at the top, we come to the global network of warehousing and shipping—UPS, FedEx, Maersk, parts suppliers like Motion Industrials and Grainger—which holds and distributes inventory.

At the far left of the x axis, product complexity, you find companies that are not particularly reliant on product design and innovation, like law firms. These companies are also relatively insulated from immediate direct changes to their business. But as you move farther right, you encounter businesses that drive their business through innovation. This is a diverse group, from most electronics manufacturers to apparel companies that constantly generate new iterations of brands. Also in this category are producers of novel materials for uses ranging from medical implants to lightweight, high-strength products for aerospace to novel LCD technology. From Nike to Microsoft to Stryker to Airbus to Fujitsu and beyond, this is an incredibly broad category of companies that will experience a major disruption.

Find your company and which impact zone you are in. If you are in the peripheral areas, you're not off the hook. You should still educate your teams and employees. The disruption will at a minimum impact your customers, which will affect you. Take a financial services company. They have no direct impact. But they still have huge indirect exposure. Their clients often include businesses that work in supply chain or high-volume production. If you are in the lowest area, you need to brainstorm how this will impact your customers, and what you want to proactively do about it.

If you are in the middle? Say you participate in a global supply chain. Or work with parts suppliers. Here, you need to begin experimenting. Within this book, we have highlighted a dozen use cases beyond prototyping where 3D printing will begin to have a significant impact. You need to pick several and get your hands dirty. What are the impacts? When will disruption occur? How does the math work? You cannot hire McKinsey to figure this out. No one knows for sure. The future will be revealed in the trenches. Get into the trenches.

Out on the edge? Your business *is* supply chain? You *are* a parts producer? The only way to win is to lead. You can't influence the outcome without taking part in writing the script. Here's a Second Wave take on the 4Ps:

PAY ATTENTION: This seems obvious, but many people might think that UPS putting 3D printers in their local stores won't impact their business. They might think of the printing service at Lowe's as a handy way to replace that chipped dimmer knob that is out of stock. But these small changes are the first hub-and-spoke model of the elastic manufacturing cloud. Don't look at things for what they are; look at them for what they could mean or what they could become. Could they become new business models?

PARTNER: Find those companies and communities who are innovating, and aggressively seek to partner with them. For very little money, you can help them along, create allies, and learn much more than you can just following your Twitter feed. Winners in a 3D printed world will be part of mutually beneficial ecosystems. This may make for strange bedfellows at times. Why are Oreo and Twitter teaming up to make a vending machine? But it is the only way to develop a rich network.

PILOT: Pick several areas where you think there may be threat or opportunity and start experimenting. Every company can identify right now one hundred or even one thousand parts sitting in inventory that could be replaced with real-time, on-demand production. Critical objects could be dramatically redesigned using advanced new software technologies and then 3D printed. For many, 3D printing will not be pervasively disruptive for decades to come. But there are opportunities for everyone to start experimenting right now.

PICK A LANE: Once you have solid assumptions about where you think things may go, where you want them to go, and how your organization might play in this new game, then jump in—big time. This means placing bets while the CFOs scratch their heads. There is no way to know exactly what revenues and costs will be in a very new model. But if you want to lead, you need to try first. The First Wave was about building capital assets to keep others out. It was about first-mover advantage. The Second Wave is about who is seeing the most the fastest. It's about first-learner advantage.

Talking to dozens of company executives, I find that the most aggressive people in this new space seem to be those in the middle. They are impacted, but it's not life threatening. It seems to me that these executives are honestly interested in figuring out what new opportunities will be arising. Ironically, it's the companies out on the far edge that are actually the most hesitant and skeptical. They are the least likely to experiment. It's almost like they are frozen by the prospect of such a big change sweeping over them. They are saying: "This won't happen." or "It may happen, but not any time soon." So they close up. They don't want to talk about it. They dismiss others who do. This is a very dangerous strategy. Because by the time you actually see it happening, by the time you find out—it's too late! Just ask Kodak, *Newsweek,* Sears, or the global taxi industry!

"People and companies that are using 3D printing are far more competitive," said Michael Mendenhall, CMO at Flex. "Currently the applications are largely about speed, but this will eventually become a mass-market production technology. The capability that 3D printing delivers is quickly becoming very interesting for a lot of different players."

We have profiled companies such as GE, Ford, and UPS in *The Great Disruption* for several reasons. First, they are classic examples of companies that were dominant during the First Wave of manufacturing who are now trying to make the shift to a new paradigm. Their actions are indicative of what companies across all industries will be required to do. But these companies are not just reacting to these sweeping changes, they are leading the way.

"The pace of change has accelerated dramatically," says Beth Comstock, GE's vice chairman and head of global innovation. "If you sit around and watch you will be left behind. Certainly technology is changing rapidly in areas like 3D printing, but it is the convergence of numerous new technologies that will rewrite industries. It's not about predicting the rate at which individual technologies will advance. It's about anticipating how these technologies will overlap, increasing the breadth and pace of change exponentially, and creating entirely new business models."

It's true that some industries are going to take years or decades to be fundamentally transformed by 3D printing. But to get an idea of how fast the transition can happen, look at hearing aids.

The business seems like a natural fit for the advantages of 3D printing. The original process was complicated. It involved nine steps, from making the molds to doing a finishing trim on the final piece. Technicians were employed to do much of the work manually. It took over a week to create a finished product. Even then, the fit was often imperfect. The hearing aids frequently wiggled around loosely inside the customer's ear.

With 3D printing, the whole process was reduced to three

steps. First, an audiologist scans the patient's ear with a digital camera. An engineer models a piece with the custom geometries. The shell is then manufacturing on a 3D printer. The product is finished within a day. It fits ears perfectly.

We've heard this story before in many other industries. Here's the astonishing part. The hearing-aid industry converted to 100 percent 3D printing in *less than 500 days*. Of those companies that refused to change over from their traditional manufacturing process, none survived. This story is not supposed to simply scare you into running out and buying whatever 3D technology you can. But it should spur you into developing a strategic plan for the coming change in manufacturing.

Few industries will flip 100 percent to 3D printing as quickly as hearing aids, but the example is still relevant to all of us. The truth is that technological disruptions don't happen in a linear fashion. Look, for example, at the growth of Airbnb. Founded in 2008, the company started off slowly. By 2011, it had only booked a total of 500,000 rooms, or 170,000 rooms a year. But then a switch

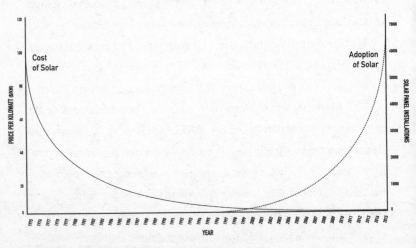

SOURCE: BLOOMBERG, EARTH POLICY INSTITUTE, TREEHUGGER

flipped. Over the next twelve months, the company booked over 9,000,000 rooms.

The same dynamic played out in consumer adoption of solar power. The price of a solar panel per watt declined significantly for three decades beginning in 1975, but these advances barely moved the needle on home installations. Then, in 2008, solar reached a tipping point. Installations have surged exponentially ever since then. Why, after years of dropping prices, did adoption take off in 2008? Because that was the year that the cost of solar *relative to other power sources* tipped. Technology adoption is like the temperature dropping from 33 °F to 32 °F. It's true that the absolute change isn't very large—only one degree. But the world suddenly looks completely different. This is why graphic representations of technological adoption look much more like hockey sticks than straight lines. 3D printing had already had decades of slow growth. Next comes the rapid vertical proliferation.

Even once you know the shape of change, predicting the exact point at which the curve suddenly accelerates is still difficult. Let's consider one transformation that 3D printing is already driving: the shift from physical to virtual inventory. It will take much longer than five hundred days for a substantial overall percentage of spare parts held in warehouses to be replaced by production on-demand using 3D printing. In fact, it's reasonable to predict that it will take, say, a decade for 3D printed virtual parts to comprise just 1 percent of all inventories. But it is also quite possible that in the year following this milestone, virtual inventory grows to 10 percent.

The critical point for leaders is that disruptions of this nature sneak up on you slowly—and then accelerate dramatically. If you haven't been paying attention at the point when things take off, you will likely be left far behind. As economist Rüdiger Dornbush said, "Things take longer to happen than you think they will, and then they happen faster than you thought they could."

CUSTOMERS TAKE THE LEAP

There are those who, even after reading this book, will still be skeptical. To those of you, I would like to introduce you to LEAP. This is GE's innovative new jet engine, the one with the new 3D printed fuel-injection system that provides 15 percent savings in fuel. It also features the lighter engine bracket designed by Arie Kurniawan in his small office in Indonesia. You may still be skeptical about the impact of industrial 3D printing, and when the future will actually arrive, but it turns out, customers are not. They are actually very excited about the future. In fact, many of them think this 3D printed future is already here.

Pre-sales for the LEAP engine have already exceeded $100 billion. That's BILLION with a 3D printed B! That makes the LEAP engine one of the most successful new product launches in GE's entire history, since Thomas Edison founded General Electric 130 years ago.

Welcome to the Great Disruption.

11

HACK ROD PRINTS THE FAST LANE

This is not merely a time of breathtaking change. It is
the moment when everyman takes back control of the
future.

—MIKE "MOUSE" MCCOY

FROM HIS EARLIEST MEMORIES, "MOUSE" McCoy wanted to race.
He was practically born in leather gloves with his hair on fire.
Mouse followed that dream full tilt. Eventually, it led him to a very
unexpected place: the middle of one of the greatest manufacturing transformations in history.

Mouse grew up in the Southern California world of proto-
extreme sports. His father surfed. His father's friends could often
be found in the garage tinkering with their hot rods. Mouse

himself started riding motorcycles when he was only four years old. Ten years later he turned pro, winning several national championships in his teens. As a young adult, Mouse used his skills to become one of the top stuntmen in Hollywood. He fell in love with the world of action photography. As Mouse entered his thirties, he decided he wanted to combine his passions for racing and film, and like every challenge previously, he wanted to do it in a big way.

The Baja 1000 is a brutal daylong race through the dusty desert terrain of Baja California. The course is an unpaved track with random ditches and jumps, and there are very few rules. In 2003, Mouse set out to record the chaotic race in a documentary. He called together his buddies from the film industry, stuntmen and camera shooters, and gave them his sales pitch.

"I had a bunch of bros," says Mouse. "I told them 'I can't offer you a lot of money. You're probably going to be up for seventy hours straight. It'll be the hardest thing you ever did, but there's plenty of beer! Who wants to go to Mexico?'" He ended up with a hundred-man crew and fifty-five cameras, all on the cheap.

Baja 1000 was a horrible environment for making a film. Given his shoestring budget, film industry people told Mouse he couldn't possibly make a documentary covering over eight hundred miles of desert. But Mouse managed to get cameras embedded on the helicopters that covered the race. Then he set off to become the first person to ever ride a motorcycle solo for the whole course. The resulting film, *Dust to Glory*, became a cult classic. "Screw stunts," thought Mouse, "I want to be a director!"

A few years later, Mouse's hand was forced. He flipped an ATV off the side of a mountain while doing a stunt in Canada. His body was wrecked, with double compound fractures in both legs. The last thing he remembered was lying bleeding in the dirt, wondering if this was the end. Airlifted to a hospital, he spent the next six months in bed. To everyone's surprise, Mouse eventually walked out of the hospital. He knew his stunt career was over, so he took the upstart ethos of *Dust to Glory* and launched a film studio called Bandito Brothers.

One of Bandito Brothers' early projects was a short Internet film to help the Navy SEALs turn around their recruiting, but Mouse's aspirations were much larger. It turned out that many of the SEALs were fans of *Dust to Glory,* so Mouse made a proposal. "We know Hollywood has been bullshitting your profession for a long time. It's time to set the record straight that Bruce Willis isn't what a Navy SEAL is. Let's go tell the real story of sacrifice and laying it down."

A SEAL captain was inspired by Mouse's message and agreed to a handshake deal. Mouse went back to Hollywood to get financial backing, but was laughed out of one studio after another. His concept—an action movie starring nonactors—didn't make sense to any studio executive. It sounded like a documentary, and no one was going to spend real money on a project that would earn a million dollars at best.

Undeterred, Mouse took every dime he and his Banditos partners had and flew down to Florida to meet up with the Navy's Group Four in the Caribbean. In Miami, they rented a yacht to serve as a drug kingpin's boat. "We hired every hot actress and model in Miami we could find, and put them on the boat to play his entourage," Mouse says.

The next day Mouse got eighteen cameras positioned on the deck of the yacht and the Navy ship as well as the small inflatables that the crew would use to infiltrate the yacht. They repeated the strike on the yacht thirty-five times in one day. A few days later Banditos was back in Hollywood with what looked like a multimillion-dollar action sequence. The studios were stunned; a bidding war followed. The eventual movie, *Act of Valor*, opened at the top of the box office in America, grossing $140 million on a $12 million budget, and becoming one of the most successful independent movies ever made. Recruiting for the Navy SEALs went through the roof.

Mouse didn't pause for a second. When Hot Wheels called, he saw a new opportunity to go all out. Back in 2005, Hot Wheels was the world's most valuable toy brand for boys. But it had grown stale and was having a hard time competing for attention with newly released interactive video games. Mouse met the vice president of design at Hot Wheels, Felix Holst. Together, they brainstormed the concept of "Hot Wheels for Real." They imagined an Area 51–type secret test complex in the desert. Inside the facility, every Hot Wheel was actually built in full scale and tested by maverick engineers and stuntmen. But Mouse and Felix were not just envisioning a series of attention-grabbing, staged commercials. That was too easy.

Instead they built massive life-sized tracks, and filmed record-setting extreme stunts. In one memorable example, Mouse's close friend and business partner, Greg "GT" Tracy, was racing a jacked-up car on an iconic Hot Wheels track. His plan was to speed down an orange track, flip up around a loop, and then race into a huge leap to the final bit of track. But at the last minute, the mechanics decided on a final adjustment that made the car look "cooler." The minor tweak, made for the camera, had a major impact on real-world performance. When GT hit the bottom of the loop, his car was thrown into nearly 7Gs—roughly the same force experienced by Apollo astronauts upon reentry. Two of GT's molars exploded. The back of the car was pushed down and began to drag, slowing the vehicle down as it barely made it to the top of the loop. GT hammered the gas, fearing the car might not be able to make the jump. He shot over the chasm, clearing the other side by less than an inch.

The Hot Wheels for Real project was a huge success; the resulting web videos were viewed millions of times. Over eighteen months, Mouse, Felix, GT, and their team set three world records. Hot Wheels sales grew more than 30 percent in a category that was flat to declining. Mattel had to have been thrilled with the results but, acting cautiously, they eventually decided it was time to shut down the project.

MAKING A NEW FUTURE

Mouse and Felix had bonded over their love of cars and design and the thrill of chasing the impossible. They didn't want this journey to end. Together they plotted their next step. Maybe they could incorporate all the new, emerging technologies for design and 3D printing? By utilizing these technologies and breaking a few rules, could a few garage-based car enthusiasts outdesign and outperform the major automakers? If so, they thought it would mark a major turning point in modern manufacturing. It would put innovation and creation back in the hands of the everyman. Their challenge was set.

Their quest led them to the San Francisco offices of Autodesk, the software company that had pioneered computer-aided design in the 1980s. Today, Autodesk produces some of the most popular software for 3D printing for both industrial and consumer uses.

"I had my head blown off starting to see what they were sketching out," says Mouse. "We were building all these incredible automobile animations for our movies—but I could see that those files were just one click away from being engineering files. They were one click away from being made!"

Mouse also met a brilliant technologist named Mickey McManus, a visiting research fellow at Autodesk. Mickey's interests included the Internet of Things, digital manufacturing, and machine learning. Mickey also loved design, cars, and what he calls "extreme super users." He and Mouse hit it off instantly. Mickey was intrigued by Banditos' "car of the future" idea. Soon afterward, he made the trip down to their studio in Los Angeles.

At Autodesk, Mouse had just seen what the amazing new 3D design software tools could do. Now Mickey turned his attention to how machines could connect, learn, and evolve. With Felix, they decided to jointly pursue a tangible project: design and build a world-class car that made use of all the evolving manufacturing technology. They figured that the traditional industry was already in disruption, attacked by billionaire inventors like Elon Musk on one side and ride sharing services on the other. They wanted to see what could happen next. In short, their goal was to run circles around the major automakers.

First, they took a classic hot rod design and added roughly two thousand sensors to gather performance data. They ran the car hard, on- and off-road, and recorded more than four billion pieces of data. This data was then fed into Autodesk's Dreamcatcher software, allowing the computer to generate thousands of potential new designs. What if, for example, they could create an entirely new chassis, 30 percent lighter but significantly stronger? They gathered other data from simulated wind tunnel testing. Previously, this process took weeks in a large chamber at a cost of tens of thousands of dollars. Now they could gather the data in minutes using only computer simulations. Felix set out to design a carbon skin fiber that also served as a sensor that would take continuous performance measurements. The digital design of their dream car was quickly taking shape.

But no one at Banditos was satisfied with anything that only existed on a screen. The possibilities of generative design were eye-opening, but they were still just bits in a computer. They wouldn't really know anything until they tested their ideas in the physical world. 3D printing was the only technology that allowed them to do this. They prototyped their designs incredibly rapidly, down to each part of the car. When they couldn't use a 3D printer in the studio, they uploaded their designs to an app on the Internet, which then automatically routed the designs to the closest UPS store with 3D printer capacity. They did A/B testing of physical products.

3D printing also meant Banditos could keep one foot in the ethos of garage experimentation. Amateur hot-rod mechanics didn't worry about creating design files and molds so they could make a million of their invention. They wanted to make just one unique vehicle for their own use. Instead of drafting tables and software, they needed something physical to work with. They spent weekends tinkering. They ran through different iterations of fixes. And what they developed was sometimes spectacular.

As Felix, Mouse, and Mickey processed all this different information, they realized how complementary computer-aided design, the Internet of Things, artificial intelligence, and 3D printing were. The Internet of Things and artificial intelligence gathered an enormous amount of previously nonexistent data that could be used to improve designs. Generative design software could process this information and produce countless design possibilities. 3D printing allowed these designs to be rapidly tested. Additionally, the complex designs created by generative processes were impossible to manufacture without 3D printing. All of the technologies were pushing forward objects that were nimble, iterative, complex, and intelligent. But 3D printing was the critical enabler that allowed the innovations to come to life in the physical world.

The three men decided to call their new vehicle the Hack Rod—a combination of "hot rod" and "hack," an innovative but officially unapproved solution. Unlike today's mass-produced cars, the Hack Rod would encourage after-factory innovations. The model was not locked to the year it was built, but was part of an evolving ecology. Mouse, Felix, and Mickey wanted to tap into the design information teeming in the world. The Hack Rod's design didn't end at the beginning of the product-development process. In fact the Hack Rod wasn't even designed in the traditional sense. It was a species meant to evolve in the wild: the pothole-filled streets and crowded highways of messy reality. It would improve design by surrendering control.

EVOLVING VEHICLES

"Design for loss of control" is a difficult concept to imagine in something as concrete as a car. Mickey described how in the future a vehicle that combined artificial intelligence, generative design, machine learning, and 3D printing might work.

"Imagine a Stingray rolling off the assembly line in 2020. I buy the car. I love the design. So I slide the update filter all the way to the right. That means it's off. I don't want anyone to screw with this car. This car can only share information with its creator and with itself to become better tuned, but I don't want anything to change the car itself.

"Then I recommend the car to a friend of mine. She loves it, too, but she positions the update filter in the middle. She wants to share her performance data with the creator and all the other cars of this model, as well as her husband's car and her garage. Soon she starts getting improvements like patches to reduce emissions. The car realizes that no one in the family ever uses the damn trunk. Over years it evolves away the trunk.

"Then a third person, a young hacker type, buys the car. He jams the filter all the way to the left. He's saying: 'I want this to share in the great petri dish of all transportation.'

"Let's say that thirty years later, the cars are being checked into the auto museum. Mine is hermetically sealed behind a firewall. It hasn't been influenced by any of the cars in the connected network. The second one comes in with moderate evolutions, including doubled fuel efficiency and zero emissions." Mickey paused.

"The third car flies in."

If this sounds impossibly fanciful for 2050, you should know that Tesla already sends out improvement patches to all of its vehicles. In 2014, a Model S caught fire after striking metal road debris. Tesla's wireless update raised the road height of all the other Model S vehicles in use around the world until new underbelly armor was installed. How much more difficult would it be for all Teslas to share data on efficiency modifications?

IN THE END, what Mouse, Felix, and Mickey found was that the world of innovation and experimentation that had inspired Porsche and Shelby to push the limits of auto design and performance in their garages was possible again today. They saw that just over the horizon is an entirely new way of making things and a sea of new products and possibilities that will accompany it.

Over the course of this book, we have looked at a number of

stories like this. We've seen a Nigerian entrepreneur 3D print a tablet. We watched a Pennsylvania man run cable down a mountain to sell televisions. And we met a Japanese scientist who pioneered human cell printing at home on his Epson. What all these stories have in common is that they are about achievable ambition.

What did Mouse do? He did not raise millions of dollars to fund his project. He did not go to the largest automakers to seek assistance with the key design and production aspects of his desired car. He set out to prove that three guys in a garage, with access to tools that exist today, can outdesign, outmanufacture, and outperform the world's largest companies. And he succeeded. With this project, Banditos planted a flag in a disruptive way of making things. Banditos showed that the future of design and production will not be written by a handful of large corporations. It will include millions of upstarts and amateurs.

This is the first lesson of the Great Disruption. We are entering a world of nearly unlimited potential. No one would expect a twenty-something engineer in Indonesia with no industrial manufacturing experience to create a breakthrough design that would startle even the CEO of GE, but he did. Few might have imagined that a stunt driver, a technologist, and a toy designer could together build a car that would make all of Detroit stand up and take notice, but that is what happened.

A skeptic looking at these stories might call them anomalies. 3D printing, after all, is still significantly less than 1 percent of total manufacturing production. But these experiments are much more than anecdotes. They are cracks in the dike. To many people, they will continue to appear as disconnected events—until the Great Disruption sweeps the wall aside.

LIVE OUT OF THE PRESENT

Unfortunately, we all face a fundamental problem in anticipating the full scope of this transformation. Humans are wired to think about the future. But in critical ways, we are wired to think about it incorrectly. We get excited when we hear about 3D printed organs or dazzling new objects that unlock breathtaking performance improvements. But then we stall. Our imagination can't supply all the pieces of the future puzzle. Instead our brain fills in the missing spaces with images from the present—with what we see and know and feel every day. We cannot conceive of a world in which we get exactly what we want, when we want it. We have a hard time picturing a world where computers can multiply the number of design possibilities by factors of one thousand. We can't imagine a product in which every 10,000th of an inch can have different performance characteristics.

Psychologists refer to this phenomenon as "presentism." It describes a tendency of leaders and organizations and individuals to preserve and defend the status quo. Were we able to visualize what is truly possible free of present constraints, we would be far more likely to pursue it willingly and aggressively. Humans have trouble estimating the real potential for enormous change. The Great Disruption requires that we reframe how we think about the world.

As leaders, we long for the change that will allow us and our organizations to achieve our potential. But our brains, conditioned by millions of years of human experience, often simulate failure, not success. In effect, we imagine the future not so we can embrace it, but so we can avoid it.

The future is certainly unpredictable and sometimes scary, but it is always inevitable. Who would have predicted that a husband and wife observing their daughter at play would spark an invention that would become so pervasively disruptive? Who could have seen design software that took a "species" of a product and simulated thousands of potential offspring? The world, as much as we want it to, does not often align with our notions about the future.

This is the second lesson of the Great Disruption. Those who will ultimately succeed will not just pursue what they can see clearly in the future. They will test a myriad of possible outcomes, and follow them wherever they may lead. The future belongs not to the great prognosticators, but to those willing to innovate and experiment.

Leaders who successfully navigate the historic shift to the Second Wave will have an unwavering belief that change is possible. We like to believe that we live in the future. But if you add up what we have seen from GE, Ford, and UPS, as well as Graphene Labs and Hack Rod, they amount to a very different conclusion about where the real future is headed. For example, supply chains are not just getting shorter—they are fundamentally shifting from inefficient, long, and linear to highly distributed, nonlinear global networks. 3D printing is not just a new way to make what we already produce today, but an entirely new toolset to imagine solutions. We can design what was never before imaginable, and manufacture what was never before producible. The Internet of Things is not about simply adding sensors to things so that we can record historic data like an accountant tallies numbers. Enabled by 4D printing, this data can be used in real time to reconfigure the structure of any object for optimal performance. This is why change is so unpredictable. To every industry that will be impacted, this change can be scary.

But if there is adversity and instability in the world of the Great Disruption, there is optimism as well. Merely by printing objects in layers, we can unlock previously unimaginable complexity. By simply shifting from massive supply chains to local distributed ones, we can dramatically reduce environmental waste and damage. By linking our design visions with powerful computers, we can produce thousands of new, better alternatives to solve our biggest challenges. By putting production in the hands of everyman, we can unleash the innovation potential of our entire humanity.

In the end, the Great Disruption is an enduring testimony to the power of human innovation. Look around at the world in which you live. It may seem like the future. It is not. Right now, our ability to print objects in layers is unleashing one of the greatest transformations in human history. The Great Disruption will wash over the global landscape of production until it is virtually unrecognizable. It will usher in a new Renaissance of innovation that will reconfigure nearly every aspect of the world that we know. It will create objects that interact with everyone and everything in ways that are alien to us now. Once you begin to see our entire future printed in 3D, you can't un-see it.

3D printing is a simple technology that will reshape the world more significantly than the Industrial Revolution did over the last three hundred years. It is early, but it is inevitable. And the initial changes are already lapping on our feet.

How will you begin to print the future?

ACKNOWLEDGMENTS

IN 2003, I COAUTHORED MY first book, *The 5 Patterns of Extraordinary Careers,* which leveraged extensive research to reveal the specific behaviors that separate middle managers from those who make it to the very top of organizations. For my second book, *The Leap,* my curiosity led me to ask a broader question, can ordinary lives be transformed from good to great, and are there controllable actions that can make this leap possible? In both cases, the research led me to the same conclusion: Yes—we are more in control of changes in our careers and our lives than most of us realize. Now, with *The Great Disruption*, Mitch Free and I attempt to answer an even more critical question: Can leaders and entire organizations successfully anticipate and manage in the wake of an epochal, world-changing transformation? My motivation for each of these projects has been insatiable curiosity, and I wish to thank all those who helped me along my path of exploration.

In many ways, writing a book is like making a movie. There are a few names that get top billing, but equally important is an entire team behind the scenes that makes it all possible. At the top of the list would be our main research and writing partner, Nathan Means, whose ability to take our concepts, stories, and anecdotes and turn them into thoroughly compelling verse is unmatched in our experience. His partner (and father), Howard Means, rounded out the team with exceptional editorial direction and direct contribution. We could not have created this book without them.

We would like to thank Tom Dunne, Peter Joseph, Melanie Fried, and the entire team at St. Martin's Press. They believed in this project from the very beginning and have provided incredible guidance, direction, and support along the way. We would like to thank our agent, Al Zuckerman and Writers House, for their encouragement and excellent professional assistance. Al has been a beacon of inspiration since I met him nearly a decade ago.

We wish to acknowledge our key team members at Fast Radius, who shared this voyage of discovery with us. Specifically, Jim Niekamp, Corrigan Nolan, David Sample, Stephen Taylor, Nick Tonini, David Crowder, Emree Chapman, Gary Fudge, Anthony Graves, and our extended partners, Gloria Griessman, Miro Pastrnek, Katie Hart, Justin Epstein, Michael Horten, Ric Bailey,

Dan Beeson, and Mike Seckler. We could not have successfully made this journey without critical support from our partners at UPS, including Alan Gershenhorn, Charlie Covert, Jon Willem Breen, Alan Amling, and Rimas Kapeskas. Together, these are the thought leaders who are ushering in our amazing future in manufacturing and supply chain.

We are indebted to the participation and insights we received from executives, engineers, and scientists instrumental in creating this brilliant transformation in manufacturing. To each of the following people, our hope is that this book captured the best of what you have accomplished. You are truly the great innovators of our time: Beth Comstock, Greg Morris, David Joyce, Christine Furstoss, Marco Annunziata, Jenna Pelkey, Scott Crump, Jeff Hanson, Hector Dalton, Harold Sears, Gil Perez, Michael Mendenhall, Daniel Stolayrov, Elena Polyakova, Mike Guyer, Carl Bass, Noah Fram-Schwartz, Joris Laarman, Jennifer Lewis, Skylar Tibbits, Steven Rice, Mike "Mouse" McCoy, Felix Holst, Greg "GT" Tracy, and Mickey McManus.

We are particularly fortunate to have had access, interaction, and in many cases friendship with an exceptional group of leaders and innovators who have greatly shaped our thinking over the years. For their inspiration, candor, and guidance we would like to thank Jeff Bezos, Elon Musk, Robert Redford, General Wesley Clark, A. G. Lafley, Bono, Jack Welch, Anne Mulcahey, Francis Ford Coppola, General Stanley McChrystal, Jim Kilts, Herb Kelleher, Carlos Gutierrez, Martha Stewart, Larry Bossidy, Ram Charan, Roger Penske, Ron Sugar, Jim Collins, John Lundgren, Bob Nardelli, Tony Robbins, Jon Hirschtick, Tim Ferris, Marcus Buckingham, Dr. Mehmet Oz, Avi Reichental, Angel Mendez, Wayne Cooper, Tom Vice, Scott Miller, Daniel Casse, David Niles, Joel Schleicher, Larry Handen, Jessica Mah, Robert

Bean, John McEleney, Judy Verses, Michael Linton, Andrea Ragnetti, Sandy Ogg, Joe Pine, Robert Wolcott, Peter Sims, Keith Ferrazzi, Jim Citrin, Adele Ratcliff, Chuck Hull, Jeffrey Merrihue, Chris Peters, Roger Fransecky, David Wilkie, Carol Seymour, Jenn Cooper, David Kidder, Anne Berkowich, Sam Pettway, and Michael Dunn.

Finally, we are extremely grateful for the support of our loving wives, Dr. Lori Smith and Shirene Free. To say that writing a book is a labor of love is often an understatement. But to do so while simultaneously launching a rapidly growing start-up, securing partnerships, and acquiring companies is more accurately described as an act of lunacy. After years of marriage, Lori and Shirene continue to put up with our unrestrained ambitions and the resulting entrepreneurial roller coasters. They are the stability in our storm, and the grace that brings true meaning to any accomplishment. For that, most of all, we are forever grateful.

Rick Smith and Mitch Free
Atlanta, Georgia
May 1, 2016

INDEX

Rick Smith Can Inspire Your Organization

From innovation and entrepreneurship to specific disruptive technologies, Rick has shared his compelling insights and powerful ideas all over the world. Discover how Rick Smith can positively affect your organization's potential.

Rick is speaking on the following topics

Our 3D Printed Future: How to Position Yourself and Your Organization to Successfully Navigate This Historic Shift

Rick uncovers the truth behind 3D printing and how organizations and leaders can successfully navigate this historic shift in manufacturing and global supply chains.

Leadership in Disruption

Using examples from the most impactful technology disruptions in our past and our future, Rick creates a compelling vision of the future of business, and what leaders can and should be doing right now to succeed.

The Printable Cure: How 3D Printing Will Transform Healthcare

3D printing is going to have a massive impact on healthcare. In this must-hear discussion, Smith explores how 3D printing and personalized medicine will revolutionize healthcare as we know it.